IET RENEWABLE ENERGY SERIES 11

Cogeneration

Volumes in the IET Power and Energy series:

Cogeneration
A user's guide

David Flin

The Institution of Engineering and Technology

Published by The Institution of Engineering and Technology, London, United Kingdom

© 2010 The Institution of Engineering and Technology

First published 2010

The Institution of Engineering and Technology
Michael Faraday House
Six Hills Way, Stevenage
Herts, SG1 2AY, United Kingdom

www.theiet.org

British Library Cataloguing in Publication Data
A catalogue record for this product is available from the British Library

ISBN 978-0-86341-738-2 (paperback)
ISBN 978-1-84919-104-3 (PDF)

Typeset in India by Macmillan Publishing Solutions

Contents

Chapter 1

Introduction

Energy generation is one of the largest sources of CO_2 emissions. If the world is serious about reducing the amount of CO_2 that is being pumped into the atmosphere and if it is also unwilling to reduce its demand for energy, then it has to reduce the amount of CO_2 produced in generating that energy.

In the UK, power generation produces more CO_2 than any other source, with about 35 per cent of the UK's CO_2 emissions coming from power generation. Transport is in second place, producing about 20 per cent. Other countries have similar figures. For example, power generation in the USA accounts for 24 per cent of all CO_2 emissions. According to the Commonwealth Scientific and Industrial Research Organisation (CSIRO), every kilowatt hour of electricity that is generated from coal-fired centralised electricity generation plants produces 1 kg of CO_2.

Different types of fuel used to generate electricity result in different levels of CO_2 emissions. Table 1.1 gives an indication of the quantity of CO_2 emitted by different fuels for every gigawatt hour of electricity supplied.

Clearly, different countries have different proportions of fuel being used to supply electricity, and consequently have different proportions of CO_2 emissions from all electricity generation.

Many countries have set targets to significantly cut CO_2 emissions. The Kyoto Protocol stipulated that the targets for reduction in CO_2 levels from the 1990 levels, to be achieved by 2008–2012, should be as detailed in Table 1.2.

Table 1.1 CO_2 emissions from different fuel sources used in electricity generation (UK)

Fuel	Emissions (tonnes of CO_2 per gigawatt hour electricity supplied)
Coal	242.9
Oil	166.0
Gas	101.6
All fossil fuels	165.6
All fuels (including nuclear and renewables)	124.1

Table 1.2 CO_2 reduction targets specified in the Kyoto Protocol

Country	Target (%)
EU-15, Bulgaria, Czech Republic, Estonia, Latvia, Liechtenstein, Lithuania, Monaco, Romania, Slovakia, Slovenia, Switzerland	−8
USA (although the USA has stated that it will not ratify the protocol)	−7
Canada, Hungary, Japan, Poland	−6
Croatia	−5
New Zealand, Russian Federation, Ukraine	0
Norway	1
Australia	8
Iceland	10

It is therefore critically important to cut CO_2 emissions from power generation if reduction targets are to be met. There are various methods and strategies that can be adopted to do this:

- using renewable energies that generate electricity with a minimum of emissions;
- switching from high to lower CO_2-emitting fuels (such as replacing coal with gas);
- using carbon sequestration, which collects and stores CO_2 to prevent it from entering the atmosphere;
- using energy conservation, which reduces the energy required to produce the effect; customers buy energy for what it can do rather than for the energy itself;
- using cogeneration, sometimes called combined heat and power (CHP), which improves the efficiency of energy produced. As a result, the use of cogeneration means that less fuel is used, and therefore fewer emissions produced, in generating the same amount of energy. Cogeneration produces more energy from less fuel.

There are other concerns in designing energy policy, such as reliability of supply, diversification of fuels and improvement in the transmission infrastructure.

Cogeneration is an available and proven technology that can be used to go some way towards meeting these needs. Both the US Federal Government and the European Union (EU) have said that they are committed to doubling cogeneration use by 2010. Within Europe, Denmark, Finland and the Netherlands generate over 33 per cent of their electricity needs from cogeneration, while Austria, Germany, Italy, Portugal and Spain generate over 10 per cent of their electricity needs from cogeneration. However, the average amount of electricity generated by cogeneration across the EU as a whole is less than 9 per cent, indicating that there is considerable untapped potential in many countries.

Studies carried out by the UK government have shown that cogeneration is the most cost-effective way of reducing carbon emissions. Many UK industries have used large-scale cogeneration for some considerable time. More recently, house and office builders have started to make use of smaller, packaged cogeneration units. Domestic cogeneration units are being developed, and should these gain in popularity, cogeneration manufacturers will become very active.

Cogeneration uses fuel in a highly efficient manner. For example, typical, large, conventional fossil fuel power plants operate at efficiencies of 35–40 per cent. This means that most of the energy of the fuel is wasted. By contrast, cogeneration plants have typical operating efficiencies between 70 and 90 per cent. Cogeneration plants produce more output for the same level of emissions, by using the heat generated by conventional fossil fuel plants, which is normally wasted, reducing energy demand elsewhere. Alternatively, cogeneration plants will generate lower emission levels while generated the same output as that generated by equivalent conventional fossil fuel plants. In addition, cogeneration plants are usually smaller than conventional power plants with a similar output, and are usually located much closer to where the output is required. This last factor minimises transmission losses. On average, global energy loss due to transmission of electricity over long distances is equal to 12 per cent of the final consumption. Cogeneration can reduce this aspect of energy loss to a fraction of its current level because most cogeneration plants have no need to transmit energy over any significant distance. Obviously, large central plants have some economies of scale that can offset some of these transmission losses. However, the use of cogeneration plants will decrease the level of emissions over most centralised systems.

Aside from its role in decreasing emission levels, cogeneration is a good technology to enable and encourage distributed generation, where heat and power are generated on the same site where they are consumed. This allows users to increase the security of their energy supplies by enabling them to meet their demand during periods when grid supply is interrupted.

The British government has set a target of having 10 GW of installed cogeneration capacity by 2010, and the EU cogeneration directives came into force in 2006 to encourage the growth of cogeneration in a much wider market.

We all know that it is no longer a viable option to squander the limited supplies of fossil fuels by burning them at efficiencies which mean that more waste heat than useful energy is produced, especially as the technology exists to end this wastage easily. Legislation is finally catching up with this reality, with the introduction of the climate change levy and the EU emission trading scheme (ETS). These are designed to encourage energy users to switch to cleaner, greener and more sustainable energy supplies. Cogeneration is an excellent example of a cost-effective method of achieving this.

The UK government has set itself very ambitious targets for cutting CO_2 emissions: 20 per cent of 1990 levels by 2010 and 60 per cent by 2050. Achieving these targets will involve significant efforts and the use of a range of solutions and technologies. Cogeneration will be a key technology among those on offer; it is readily available, cost-effective and widely applicable.

Chapter 2

What is cogeneration?

Cogeneration systems generate electricity and thermal energy, and sometimes mechanical energy as well, in a single, integrated system. This contrasts with the common practice of generating electricity at a central power station and using on-site heating and cooling equipment to meet non-electric energy requirements. Cogeneration refers to the simultaneous production of heat and electricity at the point of use. The heat may be used directly for heating, producing process steam, cooling or a combination of some or all of these. Cogeneration is a proven technology that has been around for over 100 years. The first commercial power plant in the USA was a cogeneration plant designed and built by Edison in 1882 in New York.

In the early days of the development of the power industry, energy was generated at the point of use, be it at the factory, mill or mine. Excess heat produced from the generation of electricity was usually used. This could be through using the heat in industrial processes or for space heating. This was the first form of cogeneration, although the term had not been developed at that time.

In traditional centralised power plants, the heat energy of the hot steam or gas used to drive the turbines that generate electricity is not easy to use after it emerges from the outlet. In most instances, it is simply discarded. Gas-fired combined-cycle plants use the heat from the exhaust of the gas turbine to heat water to generate steam that can be used to drive a steam turbine to generate more electricity. This enables a proportion of the waste heat to be recovered.

Cogeneration captures the heat energy that would otherwise be rejected in traditional separate generation of energy forms and uses it directly so that there is little waste while converting energy from one form to another. As a result, the total efficiency of these integrated systems is much greater than that of separate systems, and in situations where both heat energy and electrical energy are required, cogeneration is an option that should be seriously considered.

Cogeneration is not a specific technology, but is an application of technologies to meet end-user needs for combinations of energy supplies. There have been a number of technology developments that have enabled new cogeneration system configurations and made a wider range of applications cost-effective. These include higher gas temperatures in gas turbines, resulting in higher exhaust temperatures and hence greater heat energy available for cogeneration; improvement in the ability to extract energy from waste heat; better materials that enable units to be smaller; and the development of new fuel sources such as fuel cells. These are covered in greater detail elsewhere in this book.

Competitive pressures to cut costs and reduce emissions of air pollutants and greenhouse gases are driving owners and operators of industrial and commercial facilities to look for ways to use energy more efficiently, and cogeneration is one cost-effective way of achieving this. Other methods of using energy more efficiently are outside the scope of this book, but include

- increasing the efficiency of the generation system,
- increasing the energy density of the fuel,
- reducing losses in transmission of energy.

The concept of cogeneration is not a new one. Before an extensive network of power lines was developed, many industries made use of cogeneration plants. As transmission grid networks developed, central power plants became the order of the day; distributed cogeneration plants became less common, as electricity from central utilities was more economic; and power plants became larger and more centralised. However, many industries still had to generate process heat on the site. Since the oil crises of the 1970s, however, the cost of fuel has tended to rise, reducing the advantages of economies of scale offered by large centralised power plants and cogeneration has become more economically attractive.

Conventional electricity generation is inherently inefficient, converting only about a third of a fuel's potential energy into usable energy. This significant increase in efficiency results in lower fuel consumption and hence reduced emissions. Cogeneration is an economically effective way of reducing air pollutants through pollution prevention; traditional pollution control is achieved through flue gas treatment, providing no profitable output and reducing efficiency and energy output.

Cogeneration systems produce two or more usable outputs: electricity, space heating, hot water or steam, chilling systems and mechanical power are all possibilities. The most typical combinations are electricity and space heating or electricity and hot water/steam production. These combinations of cogeneration are known as combined heat and power (CHP).

The most efficient cogeneration systems (exceeding 80 per cent overall efficiency) are those that satisfy a large thermal demand while producing relatively less power. As the required temperature of the recovered energy increases, the ratio of electricity to heat output decreases. The decreased output of electricity is important to the economics of CHP because moving excess electricity to market is technically easier than is the case with moving excess thermal energy.

Conventional thermal electricity-only power plants burn fuel to either convert water into steam to drive a steam turbine or burn gas to make it expand to drive a gas turbine. Most thermal steam plants are 35–40 per cent efficient, and gas turbine power plants are 40–55 per cent efficient. It has been said that such power plants are better described as waste heat plants that produce electricity as a by-product. It is certainly the case that in most conventional power plants, more energy goes to waste than production of useful electricity.

By contrast, cogeneration plants operate at total energy efficiencies of 75–95 per cent, which means that almost all of the fuel is put to productive use. This is because the outlet heat created from generating electricity is used as heat energy, and as a result, there is little waste and less fuel is required to generate the same amount of useful work as separate generation of electricity via central power plants and heating through boilers. Because less fuel is used, the plant owners receive a number of obvious benefits:

● reduced fuel costs;
● reduced fuel supply needs, bringing about a reduction in the logistical needs to transport and store the fuel;
● reduction in emission levels;
● fewer pollutants passing through the engine or turbine, resulting in a reduction in wear.

A typical cogeneration plant consists of a number of basic elements, including:

● a prime mover such as an engine, a gas turbine, a steam turbine or a fuel cell to drive an electricity generator;
● an electricity generator to supply electricity;
● a recovery system for waste heat, which collects waste heat from the exhaust of the prime mover, and a heat exchanger, which ensures that this waste heat is put to direct use.

The proportion of electricity to usable heat generated by a cogeneration plant varies, and depending on the type of unit, it may be adjustable to a greater or lesser extent. The economic viability of a particular unit in a particular application will depend on circumstances, and is considered in more detail in Chapter 5.

Cogeneration systems can be broken down into three basic size categories: small systems that are less than 50 kW in output, suitable for single-premise domestic houses; medium systems that are 50–500 kW, suitable for medium enterprises (hospitals, hotels, flats, restaurants and leisure centres) for a wide range of applications; and large systems that are greater than 500 kW and are suitable for use in large industrial applications.

A summary of the suitability of differing methods of generating power for different conditions is provided in Table 2.1. A wide variety of fuels can be used in cogeneration systems, including natural gas, diesel, petrol, biofuels, coal, municipal waste, wind and solar energy. Cogeneration based on gas

Table 2.1 Overall summary of cogeneration systems

	Heat and power stations		Block-type thermal power stations		
System type	Heat and power station with steam turbine	Combined-cycle station with gas turbine	Block-type thermal power station with gas turbine	Block-type thermal power station with industrial engine	Micro-scale cogeneration unit with car engine
Driving system	Steam turbine	Gas and steam turbine(s) combined	Gas turbine	Industrial Otto engine with three-way catalytic converter, lean-mix engine or diesel engine with SCR catalytic converter	
Fuel	Coal, heavy oil, natural gas, heating oil	Natural gas, fluid gas, light heating oil, gasified coal	Natural gas, fluid gas, light heating oil, gasified coal	Natural gas, fluid gas, biogas, light heating oil, biogenic fuels	Natural gas, fluid gas, biogas, light heating oil, biogenic fuels
Temperature	Up to 500 °C	Up to 300 °C	Up to 550 °C	Up to 100 °C	Up to 100 °C
Main fields of applications	District heating	District heating	Process heat for industry, hospitals (steam, hot water)	Local heating networks, single buildings (hospitals, big administration buildings)	Detached family house settlements, single buildings (schools, hotels, small commercial enterprises)

Table 2.1 Continued

	Heat and power stations		Block-type thermal power stations		
	5–1 000 MWe	20–100 MWe	1–10 MWe	20–1 000 kWe	5–15 kWe
Range of capacity					
Cogeneration index (power production/heat production)	0.30–0.60	0.80–1.20	0.40–0.60	0.55–0.65	0.35–0.45
Electrical efficiency	0.25–0.40	0.40–0.50	0.20–0.35	0.30–0.40	0.25–0.30
Overall efficiency	0.45–0.85	0.55–0.85	0.75–0.85	0.85–0.90	0.85–0.90
Advantages	Waste heat recovery at huge power stations	Low investment costs, high cogeneration index	High temperature level, process heat	Small size, compact construction, high overall efficiency	Small size, compact construction, high overall efficiency

engines, gas turbines and waste heat boilers that use natural gas is quickly replacing plants using coal to drive steam turbines. Cogeneration can also use renewable fuels as a power source, including waste gases from landfill sites and sewage works, solid waste from agriculture and forestry, and municipal waste.

As previously described in this chapter, cogeneration plant essentially consists of an electricity-generating system and a heat recovery system that can make use of the waste heat from the electricity-generating system. The heat recovery system can be used to return outlet heat to the electricity-generating system, create process steam, drive heating or cooling units, or provide hot water.

There are two main types of cogeneration techniques: 'topping cycle' plants and 'bottoming cycle' plants. A topping cycle generates electricity first, and is sized according to the heat demand. Facilities that generate electricity typically produce it first for their own use and then sell any of their excess power to a utility. There are four types of topping cycle cogeneration systems. The first type burns fuel in a gas turbine or diesel engine to produce power. The exhaust provides process heat or goes to a heat recovery boiler to create steam to drive the secondary steam turbine. The second type of system burns any type of fuel to produce high-pressure steam that then passes through a steam turbine to produce power. The exhaust provides low-pressure process steam; this is called a steam-turbine topping system. A third type of system burns a fuel such as natural gas, diesel, wood, gasified coal or landfill gas. The hot water from the engine jacket cooling system flows to a heat recovery boiler, where it is converted to process steam and hot water for space heating. A fourth type of system is a gas-turbine topping system. A natural gas turbine drives a generator, and the exhaust goes to a heat recovery boiler that makes process steam and process heat. A topping cycle cogeneration plant always uses some additional fuel, so there is an operating cost associated with the power production.

Bottoming cycle plants are much less common than topping cycle plants. This is because electricity can easily be bought or sold when it is in excess of site demand, whereas heat demand is usually more restrictive. As a result, the design of most of the plants is set up such that they generate the required amount of heat, with the intention of using and/or selling electricity in excess of demand.

Bottoming cycle plants exist in heavy industries such as glass- or metal-manufacturing units where very high furnace temperatures are used. A waste heat recovery boiler recaptures the waste heat from the manufacturing heat process. This waste heat is then used to produce steam that drives a steam turbine to generate electricity. Since fuel is first burned in the manufacturing process, no extra fuel is used to generate electricity.

Three of the main technical parameters of a cogeneration system are as follows:

- total system efficiency;
- power-to-heat ratio;
- fuel energy savings ratio.

The performance of a system depends on the load and environmental considerations. On the other hand, the degree of utilisation of the energy forms produced is affected by the initial design of the system, the cogeneration strategy and the matching between the production and use of the useful energy forms. As a result, indices over a period of time, such as annual indices, are often more important than instantaneous indices, as they reveal more about the real performance of the system.

2.1 Electricity-generating systems

There are a number of different electricity-generating systems that can be used in a cogeneration plant. These include diesel or gas engines, gas turbines, fuel cells, as well as wind turbines and other forms of renewable energy.

2.1.1 Gas and diesel engines

The reciprocating internal combustion engine has been around for a long time, as this engine forms the basis of the automobile. There are a large number of models of engine that are available, and the technology exists to produce engines with generating capacities of a few kilowatts to over 5 MW. Gas engines can run on a variety of fuels, including propane, petrol or landfill gas. Diesel engines can also be used to operate in a dual-fuel mode, burning natural gas with a small amount of diesel fuel used as pilot fuel.

Engines are cheap to buy and install, and they are easy to operate and maintain and there is unlikely to be a shortage of people skilled in maintaining these engines, at least at a basic level. Engines are quick to start up, and have good load-following characteristics, and it is possible to change the output with ease. Engines also have good heat recovery potential. Waste heat can be taken from the engine exhaust and the engine cooling system, to produce either hot water or low-pressure steam.

The reciprocating, or piston-driven, engine is a well-understood technology with a long history. It operates reliably over a wide temperature range and is easy to service and maintain. Engines need fuel, air, compression and a combustion source. Most engines use a four-stroke cycle, consisting of intake, compression, power and exhaust. During the intake stroke, the piston moves down the cylinder as the intake valve opens and the upper portion of the cylinder fills with fuel and air. The piston returns upwards during the compression cycle, compressing the fuel–air mixture. When the piston is near the top of the cylinder, the spark plug emits a spark to ignite the fuel–air mixture, producing a rapid expansion of the gas mixture, forcing the piston down during the power phase, turning the crankshaft and producing power. As the piston rises up the cylinder again during the exhaust phase, the exhaust valve is opened, allowing the exhaust gas to be expelled from the cylinder (Figure 2.1).

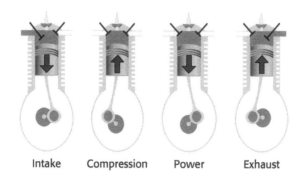

Intake   Compression   Power   Exhaust

Figure 2.1   Step-by-step valve piston operation [illustration courtesy US DoE]

2.1.2   Gas turbines

Gas turbines are available in outputs ranging from 500 kW to 250 MW. They have very hot exhaust temperatures, and can enable production of high-temperature process steam. Industrial processes can make use of this steam directly for heating or drying purposes.

Gas turbines are very useful for cogeneration in industrial applications, especially in circumstances where the industrial process produces a burnable fuel as a waste product. For example, a refinery might use by-products of the refining process to power a simple-cycle gas turbine to provide base-load power for the plant. A heat recovery steam generator (HRSG) can be used on the exhaust to produce steam for process use within the plant.

A gas turbine burns fuel to expand air by heating it. The movement of this air as it expands spins the turbine. Gas turbines have three main parts:

- a compressor to compress the incoming air to high pressure;
- a combustion area to burn the fuel and produce high-pressure, high-velocity gas;
- a turbine to extract the energy from the gas flowing out of the combustion chamber.

Figure 2.2 shows the general layout of an axial-flow gas turbine.

In this turbine, air is sucked in from the right by the compressor. The compressor is basically a cone-shaped cylinder with small fan blades attached in rows. Assuming that the light-gray area in the figure represents air at normal air pressure, as the air is forced through the compression stage, its pressure and velocity rise significantly; in some cases, the pressure of the air can rise by a factor of 30. The high-pressure air produced by the compressor is shown in dark gray in the figure.

This high-pressure air then enters the combustion area, where a ring of fuel injectors inject a steady stream of fuel. The main design difficulty in this section is to keep a flame burning continuously. This is achieved through the use of a

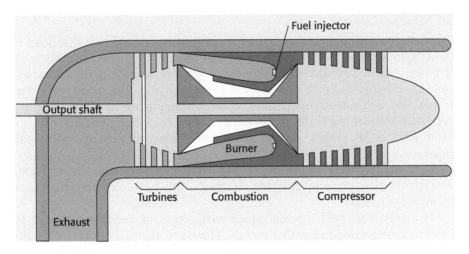

Figure 2.2   Axial-flow gas turbine layout

can, as shown in Figure 2.3. The can is a hollow, perforated piece of heavy metal.

The injectors are on the right, and compressed air enters through the perforations, with exhaust gas exiting on the left. A second set of cylinders wraps around the inside and the outside of the can, guiding the compressed intake air into the perforations. The turbine section is to the left. There are generally two sets of turbines, one to directly drive the compressor and the other to generate output power.

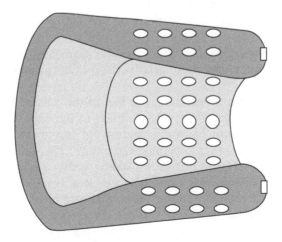

Figure 2.3   Can in cross-section

2.1.3   Fuel cells

Fuel cells are still in an early stage of development, and are best described as giant batteries that store electric potential rather than a traditional form of power generation, where fuel is burnt to generate electricity for immediate use. Like batteries, fuel cells produce electricity in the form of direct current through an electrochemical process without burning fuel. The big difference between fuel cells and batteries is that the latter can deliver only a finite amount of stored energy, while fuel cells can generate electricity indefinitely.

Fuel cells operate by means of two electrodes that pass charged ions through an electrolyte to generate electricity and heat, using a catalyst to enhance the process.

However, fuel cells are still at an early stage of development and an expensive method of generating energy. There are also issues over reliability because of the lack of experience of operating the technology, and there is a lack of the necessary support infrastructure as well. These result in fuel cells being a risky option. However, fuel cells have many advantages, and may well become a preferred option in the near future.

Operating temperatures for fuel cells range from ambient to 980 °C, with electricity-generating efficiencies of 35–50 per cent. Fuel cells are also quiet and emission free, which can be an important consideration in crowded urban environments.

There are a number of different types of fuel cells. However, all operate on the same basic principles. Layers of materials with distinct electrochemical properties are sandwiched together to form a single galvanic cell. The centre of the fuel cell consists of a membrane that can be crossed only by charged molecules. Gas-permeable electrodes coated with a catalyst stick to this membrane, adding a layer to each side. These electrodes are in turn connected to a device that completes an electric circuit (Figure 2.4).

One of the most popular versions of fuel cell is the proton exchange membrane (PEM) fuel cell (Figure 2.5). According to the US Department of Energy, 'These cells are the best candidates for many applications, because of their low-temperature operation, relative tolerance to impurities, and high power density.'

Hydrogen flows into channels on one face of the cell and migrates through that electrode, and oxygen flows into channels to the opposite electrode. The hydrogen is oxidised into hydrogen protons, giving up its electrons to the neighbouring electrode, thus becoming the anode. This build-up of negative charge then follows the path of least resistance to the cathode.

However, this would not continue without a complete electrochemical cycle. As the electric current begins to flow, hydrogen protons pass through the membrane from the anode to the cathode. When the electrons return from doing work, they react with the oxygen and hydrogen protons at the cathode to form water, producing heat as a side effect. This thermal energy can be used

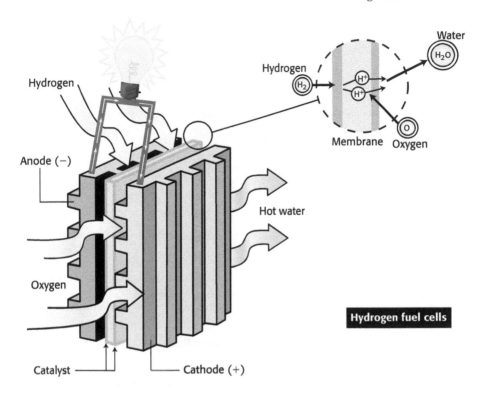

Figure 2.4 Hydrogen fuel cell

outside the fuel cell, and has been used in a number of projects to provide the heating aspect of a cogeneration plant.

To summarise:

Anode reaction: $H_2 \rightarrow 2H^+ + 2e^-$

Cathode reaction: $\frac{1}{2}O_2 + 2H^+ + 2e^- \rightarrow H_2O$

2.1.4 Micro-turbines

Micro-turbines are small units that can be installed in homes, and typically produce 30–300 kW. A heat exchanger is used to transfer heat energy from the exhaust gas to a hot water system, which can be heated to 200–315 °C.

Micro-turbines are not yet economically viable. This is because the cost of producing a micro-turbine is at present high, and the fuel cost savings to the consumer result in a payback period that is not economically viable. It is probable that the production cost will fall as economies of scale take effect. There is a very large potential market for domestic turbines. It has been predicted that these micro-turbines will achieve mass-market scale in Europe by

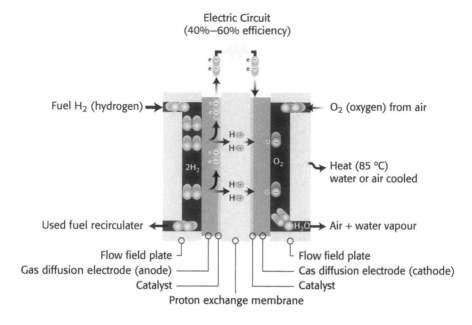

Electric Circuit
(40%–60% efficiency)

Fuel H₂ (hydrogen) →

O₂ (oxygen) from air

2H₂

O₂

Heat (85 ºC)
water or air cooled

Used fuel recirculater ←

H₂O → Air + water vapour

Flow field plate
Gas diffusion electrode (anode)
Catalyst
Proton exchange membrane

Flow field plate
Cas diffusion electrode (cathode)
Catalyst

Figure 2.5   PEM fuel cell [illustration courtesy Ballard Power Systems]

2015, with some estimates suggesting that revenues of €2 billion per year could
be achieved.

2.2   Heat recovery systems

The second part of a cogeneration system is to recover and use the exhaust heat
energy from the prime mover of the electricity system. The essence of successful
cogeneration is the beneficial use of the heat produced as a by-product of
generating electricity. The heat is contained in the exhaust gases from a prime
mover, or in the cooling systems. In the most straightforward cases, the heat
from the prime mover is used directly, without conversion to steam or hot
water. Examples include the use of hot water from cooling systems for heating purposes or the use of exhaust gases for drying. An example of this latter
case can be seen in the agricultural and food-processing industries, where
drying damp organic matter can be a time-consuming process without
assistance.

However, direct use of exhaust gases involves contact with the material to
be heated, which may cause damage. This can be through corrosion, erosion or
a combination of these, and these factors are generally more significant at
higher temperatures. The efficiency of a turbine is increased with higher inlet
temperatures, and higher inlet temperatures lead to higher temperatures
throughout the turbine. As a result, corrosion and erosion are becoming more

significant factors, especially when non-premium fuels with a high content of extraneous material are used.

Similarly, while engine cooling water can, in theory, be used directly in applications such as space heating, it is desirable in practice for cooling circuits to be self-contained and to include additives to avoid scaling and corrosion. Therefore, heat from the engine cooling water is transferred by heat exchangers to separate heating water circuits. This reduces efficiency very slightly, but it extends the life of the heating circuits significantly.

A heat exchanger is designed to transfer heat between two working fluids. The primary heat source is used to heat the secondary system. Heat exchangers can come in many forms:

- shell and tube
- plate
- cooling coil

A heat exchanger uses the fact that heat transfer occurs when there is a difference in temperature. A heat exchanger has a cold stream and a hot stream, which are separated by a thin, solid wall, and heat flows from the hot stream to the cold stream. The wall must be thin and conductive to enable maximum heat transfer to take place. However, the wall also has to be strong enough to withstand any pressure from the fluid (Figure 2.6).

This flow arrangement is called co-current. If the direction of one of the stream is reversed, the arrangement is called counter-current flow.

Figure 2.7 shows temperature profiles along the heat exchanger for both co-current and counter-current flows.

The area between the curves in Figure 2.7 is the heat transfer rate (Q). The heat transfer rate for counter-current flow is larger than the rate for co-current flow, so counter-flow heat exchangers provide more effective heat transfer.

The performance of a heat exchanger is based on the following variables:

- heat transfer area
- fluid flow velocity
- temperature gradient

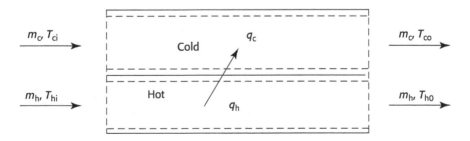

Figure 2.6   Simple flow diagram showing heat exchange

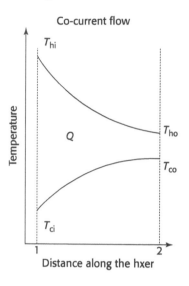
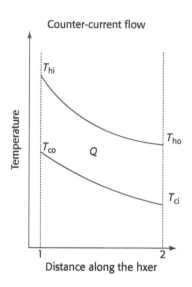

Figure 2.7 T_{ci}=Cold fluid inlet temperature, T_{co}=Cold fluid outlet temperature, T_{hi}=Hot fluid inlet temperature, T_{ho}=Hot fluid outlet temperature

The heat transfer area can be increased by adding fins to the surface. This is cheap, but it can also increase fouling. The importance of fluid flow in a heat exchanger is that it changes the overall heat transfer coefficient. The temperature gradient is the driving force for heat transfer. Fluids with a greater temperature differential between the hot and cold streams result in a greater heat transfer rate. Figure 2.7 shows that co-current flow has a high driving force at the start of the flow, but it rapidly decreases as it moves along the heat exchanger. Counter-current flow provides a fairly consistent heat transfer rate along the length of the heat exchanger, performing better than co-current flow.

2.3   Heat recovery boilers

The boiler is an essential component of any industrial cogeneration installation. It recovers heat from the exhaust gases of either a gas turbine or a reciprocating engine and, in its simplest form, is a heat exchanger through which the exhaust gases pass and in which heat is transferred to the boiler feed water to raise steam.

The cooled gases then pass to the exhaust pipe or chimney and are discharged to the atmosphere. This form of heat recovery does not change the composition or constituents of the exhaust gases from the prime mover. The exhaust gases discharged from gas turbines and reciprocating engines contain significant quantities of heat, although not all of this heat can be recovered in a boiler. Typical examples are shown in Table 2.2.

Table 2.2   Heat content of exhaust gases

	Gas turbine	Reciprocating engine
Percentage of energy input contained in exhaust gases	60–70%	35–40%
Exhaust gas temperature	450–550 °C	300–450 °C

One typical feature of the heat recovery boiler, when compared with a conventional fuel-burning unit, is that its physical size is usually greater for the same boiler output. There are two main reasons for this.

1. The lower exhaust gas temperatures require a greater heat transfer area in the boiler.
2. There are limitations of the flow restriction. Excessive flow restriction in the exhaust gas stream must be avoided as this can adversely affect the operation of the turbine or engine.

As a result, heat recovery boilers have to be designed for specific applications. The usual procedure is to provide the boiler supplier with all the details of the exhaust gas flow from which the heat is to be recovered, and with the temperature and pressure conditions of the required heat output. The boiler supplier will then have to provide the quantity of heat that can be recovered and the temperature at which the gas will be discharged from the boiler.

One important aspect of a heat recovery boiler is its control of the exhaust gas input. Normal cogeneration plant operation is determined by the prime mover. The heat recovery boiler is located downstream of the turbine or engine exhaust gas outlet and therefore tends to 'get what it is given'. Although the boiler has no control over the temperature, flow rate or constituents of the exhaust gases, it must be able to operate within its design and safety limits.

In order to control the heat input to the boiler, a set of control dampers can be installed in the ductwork between the prime mover and the boiler, with a bypass duct into which exhaust gases can be diverted. This allows the heat input and output of the heat recovery boilers to be controlled, while the prime mover output remains unaltered.

The system also allows the prime mover to be started up in isolation from the boiler, and the dampers can then be used to gradually increase the heat flow into the boiler, minimising thermal stress.

With most gas turbines and with some engines, it may be necessary to incorporate a silencer unit within the exhaust system to minimise noise levels emitted from the exhaust discharge point. In some cases, the heat recovery boiler provides sufficient noise attenuation for the main exhaust gas, and a silencer is required only in the exhaust gas bypass ductwork. In other cases, where noise is a significant consideration, it may be necessary to locate the silencer immediately downstream of the prime mover.

2.4   Uses of heat output

The heat output can be used to generate process steam and hot water, and for space heating, cooling systems, air conditioning and refrigeration units. The heat can also be used to generate additional electricity, although this introduces extra inefficiencies, reducing the overall efficiency of the system. This would be appropriate if there is a significantly greater demand for electricity than for heat energy.

Being able to switch between different outputs, such that the proportion of heat energy and electricity generated can be adjusted to suit operating circumstances and demand needs, is a benefit offered by some cogeneration systems, allowing for much greater flexibility in use.

2.5   Fuels

Because cogeneration plants can use any primary system that generates both electricity and heat energy, practically any potential fuel can be used in a cogeneration system designed for that fuel. Natural gas and methane can be used in gas turbines and gas engines, hydrogen and natural gas can be used in fuel cells, solid fuel can be used in boilers to generate steam for steam turbines and renewable energy can be used to provide heating for hybrid systems, to generate hydrogen for use in fuel cells or to be converted into pulverised gas-like fuels.

As a result, cogeneration plants can be used in a wide variety of situations, and are especially useful in disposing of burnable waste material or waste gas produced from landfill or other decomposing material. As a result, cogeneration plants can be very useful at industrial process sites.

Cogeneration systems can be used as a means of disposing of waste in an environmentally friendly way. For example, in countries with high population densities, the cost of landfill of waste can be excessive, and there is a need to find alternative means of disposing of the waste. One of the more common ways of disposing this municipal waste is by incineration. Once incineration becomes a viable alternative to landfill, the heat energy produced from the incineration process can be used in a productive manner to generate electricity and heating services for the immediate locality.

2.6   Fuel supply systems

The fuel supply systems must provide the prime mover with the required quantity of fuel, at the right temperature and pressure, so that it can operate continuously. Some prime movers, particularly gas turbines, may be capable of operating on different fuels, and in this case, the fuel supply systems must be capable of changing from one fuel to the other without the plant shutting down. Thus, provision of fuel is an essential part of the infrastructure supporting the operation of a cogeneration plant.

Natural gas is the most common fuel for cogeneration plants, because of its price, availability, wide range of applications and the lower environmental impact of its exhaust gases. The supply of natural gas to a user is made by pipeline from the national distribution network. Installation of a gas-fired cogeneration plant almost always increases the site's consumption of gas, as the new plant generates both heat and power and usually operates for a large proportion of the year. The maximum rate of consumptions usually increases as well, and this often requires the uprating of an existing site gas connection.

A number of key issues have to be taken into account when considering a new gas connection.

- The anticipated annual gas consumption, which is a function of the cogeneration plant's average fuel consumption and the anticipated number of running hours, must be identified.
- The anticipated maximum rate of gas consumption must be defined. Normally, gas turbines will consume more fuel, and generate more power, at lower ambient air temperatures, so the value must reflect the maximum hourly consumption.
- The stipulated supply pressure should reflect the requirements of the prime mover. It is always more cost-effective to have gas supplied at as high a pressure as possible, since pressure-boosting equipment such as a fuel gas compressor consumes significant quantities of electricity.
- The potential routing for new supply pipework, together with the location of metering and pressure-regulating equipment, must be determined. The general preference is for the metering and regulating equipment to be close to the site boundary so that the pipeline operator owns the pipework beneath the public highway, while the pipework on the site is the responsibility of the cogeneration plant installer and owner.

For cogeneration plants operating on coal or oil, the technical issues relating to the provision of fuel are mainly those of delivery, handling and storage. Deliveries by road or rail are off-loaded into site storage facilities from where they are delivered to the cogeneration plant. Gas oil is often the back-up fuel for gas turbine installations, and the gas-oil supply system must be designed to come immediately into operation in the event of a gas supply shutdown, whether planned or unexpected.

Handling of biomass residues depends mainly on the fuel granulometry and moisture content. Coarse residues can be transformed into homogeneous mass by crushing and chipping. Reduction in the moisture content by drying represents two main advantages: increases in the fuel heating value and a decrease in the fuel losses through fermentation during storage. Suitable technologies are available in the market to cover the handling, drying and storage requirements of different types of biomass fuels.

There are specialist engineering requirements (including health and safety requirements, pollution control and Environmental awareness that would expect some form of environmental assessment for the design of a cogeneration plant) for the handling and storage of various fuels, which must be taken into consideration. In particular, provision must be made to minimise the risks of fire, spillage or escape and to contain such problems when they do arise, thereby preventing dangerous incidents or environmental damage. The quantities stored should be determined on the basis of the need to maintain site energy provisions in the event of a supply disruption caused by the weather, shortages or other events.

2.7 Applications according to prime movers

2.7.1 Reciprocating engines

Reciprocating engines are typically used in cogeneration applications where there is a substantial hot water or low-pressure steam demand. When cooling is required, the thermal output of a reciprocating engine can be used in a chiller. Reciprocating engines are widely available in a broad size range of approximately 50–5 000 kW, suitable for a wide range of facilities.

Reciprocating engines are often used in load following applications where engine power output is regulated on the basis of the electric demand of the facility. Thermal output varies accordingly. Thermal balance is achieved through supplemental heat sources such as boilers.

2.7.2 Steam turbines

Steam turbines may be used in industrial applications to drive an electric generator or equipment such as boiler feed water pumps, process pumps, air compressors and refrigeration chillers. Turbines as industrial drivers are almost always a single casing machine, either single stage or multi-stage, condensing or non-condensing, depending on steam conditions and the value of the steam. Steam turbines can operate at a single speed to drive an electric generator or operate over a speed range to drive a refrigeration compressor.

For non-condensing applications, steam is exhausted from the turbine at a temperature and pressure sufficient for the cogeneration heating application. Back-pressure turbines can operate over a wide pressure range depending on the process requirements and exhaust steam at between 5 and 150 psig. Back-pressure turbines are less efficient than condensing turbines. However, they are less expensive and do not require a surface condenser.

2.7.3 Gas turbines

Gas turbines are cost-effective in cogeneration plants for commercial and industrial applications with a base-load electric demand greater than about

3 MW. Although gas turbines can operate satisfactorily at part load, they perform best at full power in base-load operation. Gas turbines are frequently used in district steam heating systems because their high-quality thermal output can be used for most medium-pressure steam systems.

Gas turbines for cogeneration plants can be in either simple-cycle or combined-cycle configuration. Simple-cycle applications are most common in smaller installations, typically less than 25 MW. Waste heat is recovered in an HRSG to generate high- or low-pressure steam or hot water. The thermal product can be used directly or converted to chilled water with single- or double-effect absorption chillers.

2.7.4   Fuel cells

The type of fuel cell determines the temperature of the heat liberated during the process and its suitability for cogeneration applications. Low-temperature fuel cells generate a thermal product suitable for low-pressure steam and hot water cogeneration applications. High-temperature fuel cells produce high-pressure steam that can be used in combined cycles and other cogeneration process applications. Although some fuel cells can operate at part load, other designs do not permit on–off cycling and can operate only under continuous base-load conditions.

In a unique innovation, high-temperature fuel cells and gas turbines are being integrated to boost electricity-generating efficiencies. Combined-cycle plants are being evaluated for sizes up to 25 MW. The hot exhaust from the fuel cell is combusted and used to drive the gas turbine. Energy recovered from the turbine's exhaust is used in a recuperator that preheats air from the turbine's compressor section. The heated air is then directed to the fuel cell and the gas turbine. Any remaining energy from the turbine exhaust can be recovered for cogeneration.

Chapter 3

Why use cogeneration?

Why would an operator choose to use cogeneration? There are a number of reasons why this option may be considered. These advantages can be summarised under three categories:

- economic
- environmental
- security of supply

In some cases, the benefits are overlapping; for example, using waste material from an industrial process to generate power for that industrial process gives all three benefits. There is an economic benefit arising from savings from not having to buy fuel and power, there is an environmental benefit resulting from safe disposal of a waste product and there is increased security resulting from having fuel supplies coming from within the plant. Where these benefits do not overlap, the viability of a cogeneration plant can be determined on the basis of the single benefit that does apply for that particular application. This is a comparatively straightforward assessment, but it is also a fairly rare occurrence. It is much more common for all of these factors to be involved, requiring consideration of the interacting effect.

Cogeneration is an energy concept that has been in use for many years, and where the correct conditions apply, it makes good sense to use cogeneration. Cogeneration, coupled with many different technologies as the prime mover, can provide distributed power at very high efficiencies. Typical availabilities (where availability is the proportion of time that the unit is scheduled to be operational) for mature cogeneration plants are often in excess of 95 per cent, indicating a high level of reliability. However, to state the obvious, in order for it to be economically viable, the payback time for both the investment and the operation must be commercially attractive. For this to apply, a rough rule of thumb is that there has to be a demand for the simultaneous production of heat and electricity for 5 000 h a year.

If a power station is located near a market that provides a significant demand for heating, such as a hospital, hotel or housing estate, then it would be wasteful to burn fuel to create heat that is used to generate electricity (at efficiencies of 30–50 per cent), which is then transmitted (at efficiencies of 80–95 per cent) to the site where it is used to generate heat (at efficiencies of 40–70 per cent). It is much more efficient and logical to generate heat on the site where it is needed.

Even burning fuel to generate heat that is then transmitted directly is wasteful, and so the only practical solution is to generate heat on the site. Once heat is being generated on site, and if there is an on-site demand for electricity, then it is only common sense to consider whether it is cost-effective to use the heat that is being generated to also produce electricity for use on the site.

Cogeneration units can be divided into three size categories: small units defined as those smaller than 50 kW, medium units as those being 50–500 kW and large units being larger than 500 kW.

At the time of writing, 90 per cent of cogeneration plants in the UK are in energy-intensive industries, such as oil refineries and food processing facilities. These units are almost entirely large-sized units.

There is a growing market for medium-scale cogeneration, especially in three main sectors:

- hospitals
- hotels
- leisure centres

These sectors have good load profiles for both heat energy and electricity, making for a stable and predictable demand. They typically have high energy bills, and are keen to reduce the level of these bills. They require secure energy supplies, and cannot afford to have any power outages. In short, they are ideally suited to consider the use of cogeneration units.

A number of companies believe that domestic users can represent a huge potential market for micro-cogeneration units of a size suitable for use in single-family homes, especially as the public become more environmentally aware. The technology for micro-cogeneration units is not yet commercially viable, but if they become commercially viable, there are some predictions that domestic cogeneration could lead to sales and service contracts worth over £1.5 billion per year across Europe by 2010. Another feature that would assist in making micro-cogeneration more commercially viable would be the development of cheap, reliable two-way electric meters. These would enable home-owners to either buy extra electricity from the grid, or sell excess production to the grid. As a result, the homeowner can optimise the size of unit chosen and ensure a revenue stream to offset capital cost.

Most suppliers of domestic cogeneration units believe that the most cost-effective time for a householder to install a unit would be when replacing a boiler. This would reduce the impact of the high capital cost of these units. Domestic cogeneration is considered in greater detail in Chapter 8.

The main barriers to the widespread adoption of the different options available from cogeneration technology in the UK include:

- the high capital cost involved in installing community heating pipework systems;

- the difficulty in achieving the high market penetration that would enable the costs of the pipe network to be spread over the maximum number of customers;
- lower electricity prices that can be obtained from trading electricity on the market compared to the costs of self-generating electricity.

However, with regard to the low cost of electricity available from the market, it should be noted that political instabilities in the Middle East and other regions that are dominant suppliers of fuel are resulting in fluctuating fuel prices, with a general upward trend. This is feeding into electricity prices, which have been rising steadily over the past 2–3 years.

3.1   Economic benefits

Cogeneration is most cost-effective where there are simultaneous demands for both heat and electricity. It offers several economic benefits. As indicated in Chapter 2, cogeneration is approximately twice as efficient as thermal electricity generation, and therefore needs only half the fuel to produce the same amount of electricity; hence the fuel costs are roughly halved. In addition, a number of cogeneration systems are able to use different fuels. These systems are able to take advantage of fuel price variations.

For most applications, the heating requirement determines the size of a cogeneration plant. This is because electricity is relatively easy to buy and sell to or from a local utility, and as a consequence, cogeneration units are usually sized to meet the heat demands of a site. As a result, in many cases, a cogeneration scheme will produce more electricity than it needs, and the operator has a number of options as to how to use this excess electricity.

- *Increasing the operational use of electricity.* This could be through using electricity to power processes currently powered by other means, or adding additional processes to make use of the extra electricity.
- *Selling the excess to a licensed supplier.* This is the simplest way of making revenue from excess electricity. It has the lowest costs associated with it, but it also provides the lowest returns. There is little risk associated with the revenue returns.
- *Selling via private wires.* An operator buys the electricity to distribute within a specific building such as a tower block. This gives high revenue for the electricity, but also has the highest capital costs. It is a viable option only for sites with a high density of user demand.
- *Selling via the distribution network operator's (DNO) wires.* In this, the operator is acting as a supplier. There are three options: licence-exempt supply, delivery to a nominated site in the same DNO area or a different DNO area, and supply via a licensed supplier. These options offer the potential for a high sale price for the electricity, but also entail the costs associated with using the DNO's network and the costs of settlement and

market services undertaken by the licensed supplier. Undertaking this route requires high levels of perseverance and commitment.

There is a range of capital costs, operating costs and final sale price of the electricity that is associated with each of these options. Sites such as hospitals, hotels, universities and conference centres have large on-site electricity demands. Experience suggests that installation of cogeneration plants results in such sites either increasing on-site use of electricity or selling excess electricity via private wires. However, community heating has much lower electricity demands, and generally has to go across a public network to reach end users.

In the UK, cogeneration schemes that use renewable energy can increase their income through the sale of renewable obligation certificates (ROCs). An ROC represents 1 MWh of renewable electricity generated and can be sold by the renewables generator either with or separate from the electricity generated. As a result, cogeneration schemes that make use – in whole or in part – of renewable energy sources can qualify for an additional revenue stream.

Another factor that can be of some importance for larger applications is that cogeneration systems are generally quicker to install and achieve commercial operation than conventional central power plants. Consequently, the operators receive a faster return on their investment. When one considers that large power plant units can take 2 or more years to install, this faster return can be significant.

The benefit that cogeneration plants can receive from being able to sell electricity has been eroded by the fall in electricity prices in the UK. Prices have fallen in the non-domestic market, from around £0.06/kWh in 2002 to below £0.04/kWh in 2005. Furthermore, the fossil fuel levy on electricity bills was reduced from 10 per cent in 1990 to just 0.7 per cent, resulting in a reduction in the cost of electricity from competing sources, eroding the advantage offered by the extra efficiency of cogeneration. However, there is an increase in the prospect of a rise in electricity prices over the long term as a result of political instability in the Middle East, which gives rise to an increase in the price of oil and gas.

3.2   Environmental benefits

Studies have conclusively shown that cogeneration is generally the most cost-effective way of reducing carbon emissions. It is highly fuel efficient, and ensures that maximum amount of usable energy is achieved for the minimum level of emissions at minimum cost.

The fact that cogeneration reduces demand for fuel means that it reduces the demand on natural resources, and it also reduces both the economic and environmental impacts of transporting and storing the fuel.

The European Commission (EC) has said that cogeneration is the energy technology that is best able to make the biggest contribution to cutting greenhouse gas emissions. In Europe, cogeneration currently accounts for 10 per cent of all electricity production. If this rose to 20 per cent, Europe's

CO_2 emissions would fall by 100 million tonnes per year. The EU has noted that its goal is to increase the use of cogeneration across the EU as a whole to 18 per cent of electricity-generating capacity by 2010. At the time of writing, this target remains achievable.

In addition, cogeneration plants are usually smaller than conventional generating plants, and can often fit inside existing buildings and plants. This minimises the need for additional civil construction, and hence reduces both the cost of the plant and the environmental impact that it has.

Cogeneration systems can be especially useful in locations where emission control is important, either because the local air quality is poor or because the local emissions regulations are strict. Since emissions regulations are only going to become increasingly stringent, it is likely that cogeneration systems will play an increasing role in meeting emissions targets.

Cogeneration systems can also be used to improve indoor environments, especially when they are used in conjunction with dehumidifiers to provide better humidity control than conventional systems and to reduce the potential for mould and bacteria growth.

While cogeneration provides several environmental benefits by making use of waste heat and waste products, air pollution is a concern any time fossil fuels or biomass is burned. The major regulated pollutants include particulates, sulphur dioxide and nitrous oxides. Water quality, while generally a lesser concern, can also be a problem.

Some cogeneration systems, such as diesel engines, do not capture as much waste heat as other systems. Other systems may not be able to use all the thermal energy that they produce because of their location, and are therefore less efficient than they would otherwise be.

The environmental impacts of air and water pollution and waste disposal are very site specific for cogeneration. This can be a problem for some cogeneration plants because the special equipment required to meet environmental regulations, such as water treatment and air scrubbers, adds to the cost of the project. If, on the other hand, pollution control equipment is already required for the primary industrial or commercial process anyway, cogeneration can be economically attractive.

3.3 Security benefits

Cogeneration increases security of supply. The major transmission outages in 2003 in North America, Italy and London led to millions of consumers being without power for 24 h or more. Consumers without on-site power or who did not have access to emergency back-up ended up without power. Cogeneration is an excellent form of stand-alone power, because it is reliable, economic and environmentally sound.

Security of supply is crucial in many applications, such as in hospitals and at industrial plants where an interruption of the process can cause major disruption. Such applications require the availability of constant, reliable stand-alone power. This can be achieved by using a cogeneration plant as an emergency reserve, operating only when mains power is unavailable, or by having more than one cogeneration unit such that one is always operational (essential for island operation), or by using mains supply as the reserve power supply when the cogeneration unit is out of operation (through a scheduled or an unscheduled outage). The reserve power can be maintained by operating the cogeneration capacity in a number of ways:

- hot standby, operating below rated capacity, such that it generates revenue, and output can be increased at need very rapidly;
- dual-power supply capability from the cogeneration plant and the grid, where each can be used to supplement the other as required;
- multiple units, to enable the most economic number of units to be operated at any one time.

The major power failures of 2003 in North America, Italy and London all resulted from failures in the transmission grid in well-developed countries with considerable investment in the transmission infrastructure. However, with the onset of privatisation, there has been an increasing reluctance to invest sufficiently in the grid network to meet the increase in demand. Generation capacity has generally risen in line with demand, but transmission infrastructure has lagged, and margins have become slim in some areas. On-site power supply can enable consumers evade problems in the transmission network.

Because cogeneration systems are generally located at the point of energy use, they help reduce congestion on the electricity grid by removing or reducing load. As a result, cogeneration systems help to support the grid network, giving greater reliability in transmission and distribution.

Cogeneration also improves national security of fuel supply through a number of mechanisms. It ensures that fuel is used efficiently, reducing the level of fuel imports required. For example, the proportion of energy consumption in the UK sourced from natural gas has been rising steadily since 1980, and stood at over 34 per cent in 2003. In 2004, the UK became a net importer of natural gas, with much of the imported gas coming from the Interconnector pipeline connected with Belgium. Cogeneration also allows for greater diversity in fuel supply required, spreading the risk from major interruptions to fuel supplies. It also reduces the risk profile to terrorist attacks, by reducing the dependency on single-source sites and on vulnerable transmission lines. The widespread outages of 2003 in North America, Italy and London were the result of overloading on transmission line infrastructure, and cogeneration reduces overall dependency on transmission infrastructure because it can often be operated in island mode.

It can be seen, therefore, that cogeneration provides significant benefits to security of supply, which is a key objective of European energy policy, as detailed in the November 2000 Green Paper produced by the EC.

3.4 Side-effect benefits

Another benefit that is offered by cogeneration is that it can be used to produce electricity and heat as a by-product of disposal of waste by incineration. In many countries, such as Singapore and the Netherlands, space is at a premium and setting aside land in which to bury waste is an expensive luxury. An obvious solution to this problem is to incinerate the waste, reducing its volume significantly. If this solution is adopted, then the incineration is producing waste heat. Using this waste heat to do useful work is more logical than constructing a central power plant to burn fuel to make heat. As a result, using municipal waste as a fuel for cogeneration is an attractive option for densely populated urban environments.

An example of this concept is SELCHP (South East London CHP), which collects domestic waste from the southeast London area and burns it to produce steam to drive turbines to generate electricity that is sold into the grid.

Many landfill sites give off methane gas, which is a potent greenhouse gas. Burning this methane gas converts it to CO_2, which is much less potent. As a result, burning methane is good for reducing greenhouse gas levels (although curiously, it also increases the level of CO_2 emissions). In addition, methane is a flammable gas, and it is customary to flare it off to prevent its building up to dangerous levels. If methane is already being burned, then it makes sense to put that energy to productive use. As an example, the US Environmental Protection Agency is running a Landfill Methane Outreach Program, to reduce methane emissions from landfills by encouraging the recovery and use of landfill gas as an energy resource.

3.5 Trigeneration

The heat from a cogeneration system can be used for different purposes, including heating, ventilation and air conditioning (HVAC) systems for heating or cooling of buildings. It is a simple matter to couple a cogeneration unit with the heating system of a building, thus providing space heating for that building. The main difference that this creates compared to conventional systems for space heating is the equipment used to generate the heat and the thermal characteristics of the available heat. A cogeneration system can also be used to drive the HVAC system and provide cooling for the building, as well as heating, giving rise to potential improvements in the economic benefits of cogeneration. However, the adaptation of the HVAC system of the building could involve an excessive cost.

This technique is called trigeneration, in which three different forms of energy are derived from the primary energy source. This option allows the user to have greater operational flexibility at sites with demand for energy in the form of both heating and cooling. The heat recovered from cogeneration systems can be used in HVAC systems, as shown in Table 3.1.

Table 3.1   Use by HVAC systems of heat recovered from cogeneration systems

Type of heating/cooling	Cogeneration system	
	Gas turbine	IC engine
Low temperature (hot water 40–60 °C)	Yes	Yes
High temperature (hot water 80–100 °C)	Yes	Yes
Single-stage absorption chiller (low-pressure steam, 100 kPa)	Yes	Yes
Two-stage absorption chiller (high-pressure steam, 860–1 000 kPa)	Yes	Some

The trigeneration system is improved by the use of absorption chillers. In this case, the main energy consumption is heat, which can be provided integrally by the cogeneration system. However, these cogeneration systems can also use conventional vapour compression chillers, with or without the use of absorption chillers. In this case, the main consumption of energy comes from generating electrical power. If the work produced by the cogeneration system is used to generate electricity, the mechanical chillers can use this energy. The use of absorption chillers together with mechanical chillers can, in certain cases, improve the performance of the whole system.

Different kinds of chillers can be used to produce the required cooling, including mechanical chillers, absorption chillers and desiccant chillers. The coupling of the cogeneration and cooling production can be done by the energy sources that are required by the cooling system and provided by the cogeneration system. Table 3.2 presents a summary of this. The use of mechanical chillers can be coupled with the cogeneration systems, with the electricity that is produced by the cogeneration system being used to drive the compressors of the chillers.

The heat energy source that supplies the single-stage absorption chillers can be either hot water (up to 150 °C) or low-pressure steam (up to 100 kPa). The cogeneration system can provide the chillers with these heat energy sources. The heat energy source that supplies the two-stage absorption chillers can be steam at 790–830 kPa. This steam can be produced by the exhaust gas that arises from a gas-fired engine in a cogeneration system. However, this means

Table 3.2   Main energy sources in trigeneration systems

Chiller	Energy source from generation	Main output of trigeneration
Mechanical	Electricity	Thermal energy (refrigeration)
Absorption	Heat (and electricity)	Thermal energy and electricity
Desiccant	Heat (and electricity)	Thermal energy and electricity

that the cooling production depends on the gas-fired engine. As a solution to this problem, a conventional fired boiler can be installed to produce steam while the cogeneration system is out of service or to provide the peak needs. Alternatively, the cogeneration plant could consist of several units to ensure that when one is out of operation, through a scheduled or unscheduled outage, another is available to take its place.

Thus, instead of using mechanical energy, absorption chillers use heat in order to provide a working fluid (the refrigerant), which can be expanded and cooled as part of a refrigeration cycle.

3.6 Round-up

Cogeneration provides a reliable and highly efficient source of on-site power. It reduces emission levels and allows waste products to be put to productive use. It reduces dependency on a transmission network and increases security of supply.

Cogeneration can provide an improved means of supplying power and heat that is both economically and environmentally sound.

Chapter 4
Who can use cogeneration?

Cogeneration systems of varying sizes have been designed and built for many different applications in many different places. Large-scale systems can be installed either on the site of a plant or off-site. Off-site plants have to be located close enough to a steam customer to enable them to cover the cost of a steam pipeline. Industrial or commercial facility owners can operate the plants, or they may be operated by a utility. At present, about 90 per cent of cogeneration in the USA is used by industrial manufacturers.

Cogeneration systems are also available for small-scale users of electricity. Small-scale packaged or modular systems are being manufactured for commercial and light industrial applications. Modular cogeneration systems are compact and economic to manufacture. These systems range in size from 20 kW to 650 kW and produce electricity and hot water. It is usually best to size the systems to meet the hot water needs of a building. As a result, the best applications are for buildings that have a constant and continual need for hot water or steam, such as hospitals, hotels or restaurants. In these instances, cogeneration plants can be used to ensure a reliable supply of power, heating and hot water/steam, even during interruptions to mains supplies. An example of this is shown in the case study on cogeneration as hospital backup.

Case study: Cogeneration as hospital backup

Northampton General Hospital has operated a cogeneration plant since 1989 in order to ensure that it had a reliable and secure supply of power and heating. This was tested during a major incident exercise, when a simulated serious accident with a large number of 'casualties' (a regular test scenario hospitals undertake to test procedures) was undertaken.

The hospital thought it important to simulate a situation where a total boiler failure had taken place. This catastrophic boiler house failure meant that the boiler could not provide heat or hot water. The cogeneration plant supplied the hospital with its essential heat and electricity. The 'patients' continued to receive the best care and attention through the use of cogeneration as an emergency backup, and power, heating and hot water supply were not interrupted.

Several companies have attempted to develop cogeneration systems with capacities of up to 10 kW sized for use in private homes. These will be able to replace domestic boilers and provide heating and hot water, and also generate electricity for use in the house, with any excess being sold to the local utility and any shortfall being purchased from the utility. This would enable the household to make best use of its own consumption pattern, with the option of running its own supply when the price of electricity is high (and selling any excess at favourable rates to the utility) and buying electricity from the utility when the price is low, leaving the cogeneration unit switched off. A number of trials have been started, but at the time of writing, no domestic cogeneration systems were being offered commercially. Domestic cogeneration is covered in greater detail in Chapter 8.

Numerous governments around the world are committed to a significant increase in the amount of cogeneration in use in their countries. Both the US government and the EU have declared their intention to double the use of cogeneration over the current rate by 2010.

In the USA, the Department of Energy (DoE) and the Environmental Protection Agency (EPA) are trying to eliminate obstacles to the introduction of cogeneration schemes in the USA. The EPA has introduced a Combined Heat and Power Partnership, a voluntary programme to reduce the environmental impact of power generation by promoting the use of cogeneration. The partnership acts as a liaison between the CHP industry, state and local governments, and other stakeholders, supporting the development of new projects. To help encourage the development of highly efficient CHP projects, the EPA and the DoE present Energy Star CHP Awards to projects that are in commercial operation, operating within the emission limits stipulated in their permits, and have a minimum of 12 months and 5 000 hours of measured operating data. Thermal energy must comprise 20–90 per cent of the total system output.

Winners of the 2005 Energy Star CHP Awards include

- Hexion Speciality Chemicals CHP Project, Moreau Industrial Park of South Glens Falls, NY
- The CHP Project at the University of Texas at Austin
- Arrow Linen CHP Project, Brooklyn, NY
- Rego Park Nursing Home, Flushing, NY
- Hermany Farms CHP Project, Bronx, NY

Cogeneration currently produces about 10 per cent of the electricity generated in the USA.

The EU passed the European Cogeneration Directive, which came into force in February 2004. The aim of this directive is to increase energy efficiency and improve security of supply through the creation of a framework to promote and develop high-efficiency cogeneration, resulting in savings in the energy market. The directive takes into account specific national circumstances, especially the climate and economies in the different countries. The average amount of electricity generated by cogeneration across the EU-15 is currently 9 per cent, although this varies widely from country to country, with Denmark, Finland and the Netherlands all producing over 33 per cent of their electricity generation from cogeneration. Austria, Germany, Italy, Portugal and Spain each produce over 10 per cent of their electricity generation from cogeneration.

Finland has a high percentage of its electricity generated by cogeneration because it has a large number of pulp and paper mills, which are ideally suited for cogeneration, and it has a high concentration of demands for both heating and electricity, making cogeneration a logical choice. Denmark and the Netherlands both are densely populated countries, and have a tradition of concern for the environment and, as a consequence, have well-developed cogeneration industries.

The European Commission has estimated that there is considerable untapped potential and that doubling the amount of cogeneration in service by 2010 is achievable.

The UK government has undertaken a study in order to determine the most cost-effective way of reducing carbon emissions. This study identified energy conservation as being the cheapest, cleanest and safest way of making significant cuts in CO_2 emissions and meeting its Kyoto targets. The study further concluded that cogeneration was the best solution, arguing that cogeneration was a 'no regrets' technology and could usually be justified on economic grounds alone, regardless of the environmental benefits. In the March 2004 budget, the UK government reduced the VAT on domestic micro-cogeneration devices to take effect from 2005 so that these devices pay the same VAT rating as other energy-efficient devices, hoping to encourage investment in this technology. The government study estimated that micro-cogeneration could cut the typical household energy bill by £150 a year and reduce CO_2 emissions from the household by up to 1.5 tonnes per year.

4.1 Who uses cogeneration?

Medium-scale (50–500 kW) cogeneration units have had a successful track record in Europe in a wide range of building applications, but this sector is currently the smallest cogeneration sector in the USA. Sites that have a large hot water demand, such as colleges, hospitals, hotels and some restaurants, appear to be the most attractive potential markets.

There are many types of establishment that frequently combine a sufficiently large heat load with a fairly consistent electrical load. As a consequence, these types of establishment are ideally suited to consider the use of a cogeneration plant. Such sites include

- hotels
- hospitals
- university campuses and boarding schools
- office buildings
- swimming pools and leisure centres
- stores and supermarkets

This is not a complete list, and there are many other places where cogeneration might be feasible, such as army barracks, museums and prisons. As a rough rule of thumb, any establishment with a fairly reliable demand for more than about 50 kWe and 50 kWth can be considered to be potentially highly suitable for cogeneration.

4.1.1   Hotels

Hotels that accommodate over 50 bedrooms often have a significant heat demand for space heating and catering for as much as 18 hours per day for much of the year. Furthermore, the electricity demand from such hotels is fairly constant, so cogeneration units can be applied very effectively. The level of demand imposed by hotels of this size can vary according to location (which can affect the extent of space heating required) and the quality of the hotel (as a general rule, better hotels offer more services that require electricity and hence results in increased electrical demand).

4.1.2   Hospitals

Energy demand in hospitals tends to be continuously high throughout almost the full day and over most of the year. A high heat load is combined with a high electrical demand, and all electricity generated can be used on the site. Hospitals have a need for absolute reliability of supply of both electricity and heat energy.

4.1.3   University campuses

At campuses, some boarding schools and other educational institutions, thermal and electrical energy demand is usually high enough for sufficiently long periods to make the application of cogeneration suitable for this type of establishment. Universities can also make use of the expertise of their engineering students to evaluate different proposed concepts and alternatives, achieving the twin aims of reducing the cost of employing consultants to do this work and giving the students valuable experience in practical design.

4.1.4   Office buildings

A high space-heating demand in winter is often combined with cooling requirements in summer in an office building, resulting in a fair demand for heat energy to either act as space heating or drive air-conditioning units.

Offices often have very high electrical demands for lighting and using power appliances. This electrical demand can be consistently high for 10–12 hours a day, over almost the entire year, making it possible to operate a cogeneration unit for approximately 4 000 hours per year. If no cooling capacity is required, then the application of cogeneration in an office building is questionable, since in that case, insufficient operating hours can usually be achieved.

4.1.5 Swimming pools and leisure centres

Swimming pools and leisure centres have a demand for both thermal and electrical energy for 12–14 hours per day throughout the year. Thermal energy is required to heat the pool water and provide ventilation air and domestic hot water, and electricity is needed to operate pumps and lighting. Furthermore, at pool water temperatures below 30 °C, the application of a condenser in the exhaust system of the prime mover can increase the overall efficiency of the cogeneration system to over 90 per cent.

4.1.6 Stores and supermarkets

Large stores and supermarkets often have a high demand for space heating and cooling energy, while their lighting levels are high for 10–12 hours per day. Most of the energy consumption in supermarkets is attributable to the electrical loads that are required to meet sales objectives (bright lighting), preservation of foodstuffs (cold stores and refrigerated displays) and air conditioning.

Case study: Jurys Towers Hotel, Dublin

The Jurys Towers Hotel, which is located in Ballsbridge, Dublin, consists of two adjoining hotel blocks. The main part of the hotel comprises 300 bedrooms, function rooms, restaurants, lounges, pubs, a leisure centre with swimming pool and a business centre. The separate tower block contains a further 100 bedrooms. With a substantial year-round demand for electricity and heat, this hotel was considered to be an ideal location for the installation of a cogeneration unit. Prior to its installation, the heating services for the hotel were provided by two ageing steam boilers, each rated at 4 765 kg/h.

Inenco recommended the installation of a 300 kWe cogeneration plant and the replacement of the existing steam boilers with two 1 400 kg/h boilers for the laundry and two 2 200 kW low-pressure hot water boilers. Temp Technology supplied the chosen cogeneration unit. The plant is integrated into the hot water system, and the hot water is circulated via the cogeneration unit in the primary water loop before being returned to the boilers. Therefore, the boilers are only required to supply the make-up heat for raising water to the desired temperature (90 °C).

Cogeneration unit technical data:

Type: Dorman 6 DTg
Electrical Power: 304 kW
Heat output: 445 kW
Fuel Input: 999 kW

In general terms, cogeneration can be useful and economically viable for the following circumstances:

- Any industrial process or group of buildings that has a consistent and high enough demand for both heat energy and electricity.
- Any industrial process that has a constant requirement for the provision of steam or hot water.
- Any industrial process that produces burnable waste as a by-product, especially when that waste is costly to dispose of by other means.
- As a final stage in a recycling process. The classic recycling hierarchy is reuse, repair, recycle and dispose. Disposal by incineration is one option that is often chosen in locations where landfill is prohibitively expensive.
- Methane is produced as a side effect from the decay of material in a landfill site, and it is vitally important to prevent the build-up of pockets of methane. The methane is flared off, and where there is a nearby demand for heat or electricity or both, this flaring can be productively used.

Cogeneration is particularly attractive for supplying energy to district heating schemes or to groups of buildings that require both heat energy and electricity, such as hospitals, hotels or universities. Any place that has a demand for both a central supply of heat and electricity is potentially a viable site for cogeneration, and this should be considered to determine whether cogeneration would be a cost-effective solution.

Examples of potential cogeneration applications include large chemical plants and paper mills and small communal properties such as hospitals, hotels, leisure centres, nursing homes and sheltered accommodation.

While considering the viability of cogeneration for a specific application, the first aspect that needs to be taken into account is whether the application requires simultaneous need for heat and electricity over long periods. The longer the period that this situation applies, the better the economic and environmental case for cogeneration. The case is less strong if demands for heat and electricity are not simultaneous, as storage or sale of the non-required energy form has to be arranged. If simultaneous demand exists only for a short period, then the cogeneration plant may lay idle for some time.

The next stage is to determine the viability of the economic case for a particular cogeneration plant. The best way to do this is to first evaluate the heat demands of the complex and then to match the thermal output of the cogeneration unit to this in order to maximise the running hours. As far as electrical generation considerations are concerned, the cogeneration unit should meet the entire base load of the site, so that all the generated electricity can be used on the site. It is usually the case that minimising the amount of electricity exported to the grid is the most cost-effective strategy for generating electricity from a cogeneration plant. Generating less electricity than the base-load requirements involves importing electricity from the grid, often at premium rates because the time of greatest demand from a site usually coincides with the time of greatest demand on the grid. Likewise, generating more electricity than what the base load of the site requires usually occurs at a time when electricity prices are low because they tend to occur when general levels of demand for electricity are low. The key to maximising the economic viability of a cogeneration unit is to maximise the percentage of electricity that is consumed on the site. When operating in parallel with the local supply network, any electricity required above that generated by the cogeneration unit would have to be imported from the grid.

The industrial sector is the largest and best characterised of the cogeneration segments worldwide. It is also the cogeneration segment that has the greatest potential for near-term growth. Most of this capacity is in industrial sites that have large steam loads. In 1994, three industrial sectors, pulp and paper, chemicals and petroleum refining, accounted for 85 per cent of all industrial cogenerated electricity in the USA. Similar figures apply elsewhere in the world; for example, pulp and paper industry is an important sector in Scandinavia. Pulp and paper accounted for 41 per cent or 59 TWh, the chemical industry accounted for 33 per cent (47 TWh) and petroleum refining made up 10 per cent (14 TWh) of the cogenerated electricity in 1998.

The cogeneration systems in these industries typically generate an electrical output of over 25 MWe, and have steam generation rates that measure in the range of hundreds of tons of steam per hour. These plants are generally owned by an independent power producer that seeks an industrial customer for the steam, reducing the net operating costs of the plants in order to improve their competitiveness in selling electricity. Cogeneration has been applied widely in the following industry sectors:

- *Agrofood*: In the sugar industry, for example, autonomous electricity generation with steam turbines is common. There is a large potential for cogeneration in all sectors of the agrofood industry, as much of this industry produces burnable waste products (such as corn husks) and have a demand for steam and electricity.
- *Brewing*: Many breweries are too small for cogeneration to be commercially viable. Increasing the potential for cogeneration in this sector is dependent on obtaining realistic prices for electricity, which can then be sold to public networks. The best conditions are in Germany, the UK, Italy, Portugal and Spain. Cogeneration in a closed cycle in breweries is difficult because of the batch nature of the brewing process.
- *Brick and clay manufacture*: There is high potential for the application of cogeneration using gas turbines and engines in this industrial sector. This

sector has significant growth potential and limited application of cogeneration plant to date, giving it a big market potential for cogeneration development.

- *Cement manufacture*: Back pressure or gas turbine generation might have possible applications in this industrial sector. Potential applications include the use of waste heat from clinker cooling.
- *Ceramics*: Cogeneration with combined cycle and direct heat gas turbines has been successfully demonstrated for this sector in Germany. This sector has a big potential for cogeneration use.
- *Chemicals*: Cogeneration has been commonly applied in this sector. This sector has big market for improved technologies, including waste heat recovery from the industrial production processes.
- *Paper*: There is potential application of cogeneration for integrating the different energy processes involved in this industrial sector: steam supply, dryer heating and electricity generation. Back-up power supplies from the public network is normally required.
- *Textiles*: The ratio of thermal energy/electricity demand in this sector makes it an ideal sector for cogeneration based on gas turbines. A minimum power requirement of 800 kWe is needed for plants to be suitable for cogeneration.

Case study: Buchanan Hardwoods, Inc., Alabama, USA

When Buchanan Hardwoods planned its new plant in Aliceville, Alabama, USA, in 2003, cogeneration was considered as a power supply source from the start. According to Bruce Nesmith, the then plant manager the company needed a dependable and economic method of disposing of the large volume of waste that was produced by the process of producing hardwood flooring.

Each 8-hour shift generates approximately 48 tons of waste, mainly consisting of sawdust. Hauling the waste away was viewed as a non-optimal solution. There were buyers for the waste, and the price was potentially acceptable. However, the biggest issue was dependability: would these buyers always be there, and would they always want the full volume of waste that the plant produced? Cogeneration was considered to be a logical choice to assure that waste disposal would always be available and would be economically viable over the long term.

At the plant, rough hardwood planks are kiln-dried for approximately one week prior to processing into flooring. Operation of the kiln requires the supply and use of steam. The wood waste generated by the plant is used to fuel the boiler that produces the steam. However, this steam is at a high pressure, at about 19 bar. The kiln uses steam at about 0.8 bar. Such low-pressure steam is one of the outputs of the steam-driven turbine generator. The other is electricity.

The electricity generated is used to power the plant's production operations, which significantly reduces the power bill from the local utility. When the plant's demand is lower than the electrical output of the cogeneration unit, the excess capacity is sold to a neighbouring industrial unit with a high power demand. A single utility connection serves both plants, and multiple meters are in place to verify the savings that are being realised by the cogeneration unit.

The plant operation at Buchanan is relatively simple and straightforward. The saws and milling machines in the plant are connected to a vacuum system that collects the dust generated by the plant and carries it to a storage silo. The larger pieces of wood that are not handled by the vacuum system are fed through a hog that grinds them into pieces that are small enough for the boiler's fuel feed system to handle. Conveyors then feed the wood dust to the boiler. Steam from the boiler is used to drive the turbine generator. Low-pressure steam from the turbine generator is then fed to the kilns, with the excess steam being piped to condensers that reduce it to water that is then recirculated to the boiler's feedwater system.

The cogeneration system that was chosen includes a Coppus turbine; a GE generator; two air-cooled condensers manufactured by SEECO, Alabama; a motor control centre; steam piping and all of the other associated electrical and mechanical elements. The system started operating in June 1999, and has operated successfully since then.

4.2 District heating

District heating and cooling is the distribution of heating (hot water, steam, and space heating) and cooling energy transfer mediums from a central energy production source to meet the diverse thermal needs of residential, commercial and industrial users. Thermal energy needs include space heating and cooling systems for maintaining human comfort, meeting domestic hot water requirements, and manufacturing plant process heating and cooling systems. In many of the systems that have been established around the world, both district heating and district cooling have not been provided, with just one of these being installed. For example, in Europe, where moderate summer temperatures prevail, most district thermal energy systems need to provide only heating capability.

District cooling has only recently become more widespread, with the most prevalent application being in North America, where the summer temperatures can, over extended periods, reach extremes of 30–40 °C. There are a number of factors that must be taken into account when determining whether a district heating (DH) or district heating and cooling (DHC) system should be implemented in a particular community. These factors include local economic and climatic conditions, the viability of competing alternative energy supply systems, local energy production and utilisation efficiency considerations, local environmental benefits, and differing perspectives of the producer and user on the benefits of district systems.

In general terms, DHC systems can be defined as the production of heating and cooling energy at one or more sources and the subsequent distribution of the thermal energy via pipelines to 'district' users. A typical DHC system therefore comprises three subsystems:

- thermal energy generation
- thermal energy distribution
- incorporating the thermal energy at the user's location

Although varying from country to country and from city to city, certain conditions must generally prevail in order for a DHC system to be viable. Heating and cooling load densities should be relatively high. A high total heating/cooling load is desirable since improved operating efficiencies can be realised at larger facilities and economies of scale favour larger installations.

Apartment complexes, hospitals, universities, groups of office buildings and factories are all energy user candidates that meet the prerequisites that have been specified above. Many major cities around the world meet much of their heating requirements through district heating. This is particularly true in Scandinavia and Eastern Europe. DHC systems that service areas of the city that lie beyond the high-density building zones typically result when adjacent housing densities are fairly high and/or there are several inexpensive thermal energy sources including waste heat recovery from energy-to-waste facilities, large power generation plants and gas turbine combined-cycle cogeneration plants.

Without such local opportunities for DHC supply and use, citywide applications become borderline candidates. Cities that have well-developed district heating systems include Paris, Helsinki, Stockholm, Copenhagen, Moscow, New York, Boston, San Francisco, Toronto and Tokyo. In Sweden, Finland and Denmark, district heating supplies 30, 39 and 42 per cent respectively of the entire countries' heating demand.

DHC systems are usually connected to a diverse group of customers with varying load requirements. As a result, such systems must typically be able to accommodate a relatively large total heating/cooling load with potentially wide variations from season to season. Since individual customers often experience their peak loads at different times of the day, the central production plant's daily characteristic load curve tends to be smoothed out, with the peak demand significantly lowered compared to the sum of all the individual peak loads. As a result, the installed total capacity of a DHC system can be less than that of conventional decentralised systems.

Depending on the total system peak and average load requirements and the load variations that arise from day to day and season to season, DHC plants of varying complexity can and have been developed. A relatively simple DHC system might utilise a single energy production facility, possibly comprising an oil- or gas-fired boiler for heating purposes and an electrically driven centrifugal chiller for cooling purposes. Multiple units may also be selected to more efficiently meet base, intermediate and peak loads, as well as providing standby capacity and increased system reliability.

More complex DHC systems might use several different energy production facilities, such as energy from waste (normally provided from municipal, commercial and industrial waste incineration), waste heat from manufacturing plant processes, absorption chillers, heat pumps and coal-fired boilers. Other sources of heat for district heating systems include geothermal, cement kilns, biomass and solar collectors. In the case of these more complicated thermal energy production systems, the energy sources selected and the manner in which they are used depend on local fuel prices, the availability of such alternatives, the proximity of the load to such sources, environmental sensitivities and other factors.

Case study: Lutherstadt Wittenberg

Country: Germany
Location: Wittenberg
City: Lutherstadt Wittenberg
Start of operation: 1995
Primary fuel: Natural gas
Electrical output: 6 224 kWe
Thermal output: 7 948 kWth
Use of electricity: Sold to the local utility

The cogeneration plant in Wittenberg is an excellent example of the successful replacement of old-fashioned power plants by state-of-the-art cogeneration plants. Because of an increasing demand for environmentally friendly energy, the local authorities in Wittenberg decided to change from coal to natural gas as the primary energy supply for new power plants. The energy supply of a newly built district in Lutherstadt Wittenberg was achieved through the use of a new cogeneration plant manufactured by Jenbacher.

About 20 000 people are supplied with heat from the cogeneration plant through a district heating network. The electricity that is produced by the plant is fed into the grid of the local utility at a voltage level of 6.3 kV. In addition to the reduction of 22 000 tons of CO_2, the plant reduced carbon monoxide emissions by 70 per cent and SO_2 emissions by over 95 per cent. Ash emissions are also reduced to an insignificant level.

4.3 Operation and maintenance

One advantage that is offered by cogeneration is that it can turn some waste streams from industrial processes into electricity and heat, thus turning a potential problem and converting the cost of waste disposal into a benefit and a revenue stream.

However, one consequence of this is that the owners may have only limited experience in operating and maintaining power plants. Acquiring that expertise can be costly and time consuming. On the other hand, forced outages can be inconvenient and very costly.

One solution to this is to make use of one of the firms that specialise in offering operation and maintenance (O&M) services. This enables the contracting firm to make use of expertise that would otherwise be difficult for them to acquire, while the specialist firm can concentrate its expertise into this specific area, enabling it to provide this service at a reasonable cost.

Between 30 and 60 per cent of cogeneration plants in the UK have some recourse to contract energy management (CEM). This figure is similar across Europe, and there is significant growth in this market in developing countries.

Different sites can place a different emphasis on the needs of O&M, and all sites need to find an appropriate balance between maintaining efficiency, maximising operational hours, minimising emissions, disposing of waste material productively and ensuring that there is reliable and rapid start-up. The site might need flexibility in the supply levels of either heat or electricity or both in order to match varying demands, and it might need to retain the option of both selling and buying electricity to the national grid.

Outsourcing O&M removes the risk of varying costs, thus passing this risk to a company specialising in this technology. In addition to the cost considerations, there is also the issue of the growing skills shortage that is taking place in many countries. In some parts of the world, it is becoming increasingly difficult to recruit sufficient numbers of sufficiently qualified engineers. Recruiting and retaining such staff is difficult for non-specialist organisations. This situation will be exacerbated significantly if government targets to dramatically increase the use of cogeneration plants are met.

In addition to specialist staff, a lot of O&M activities at CHP plant requires the use of specialised equipment, and – especially for gas turbines – can require the support of a specially equipped workshop. Such equipment can include oil, coolant and fuel analysers. Maintaining such equipment is expensive; it can often be used to full effectiveness only by specialist operators. This equipment is of use only during scheduled maintenance outages. Such outages represent only a tiny proportion of the life cycle of the plant, and the specialist equipment held by operators who own just a few plants would largely lie idle.

According to BP Energy, the field of CEM is one that is growing steadily and also has significant growth potential in the future.

4.4   Supplementary and auxiliary firing

Supplementary firing can raise the overall heat-to-power ratio of a cogeneration plant to up to 5:1, and it offers valuable flexibility in meeting variable heat loads. It also enables the flue gas temperature to be raised to a temperature to suit higher temperature applications. The exhaust gases from a gas turbine or reciprocating engine contain around 15 per cent oxygen, and this allows supplementary firing to be carried out in the exhaust before it passes into the boiler.

Since this exhaust is already hot, supplementary firing allows a higher combustion efficiency to be achieved than conventional boilers, enabling the same boiler output to be obtained with lower fuel consumption and reduced CO_2 emissions. Efficiencies of up to 88 per cent can be achieved, which compares well with the 80 per cent efficiency that is typically associated with natural gas combustion in a conventional boiler. Because supplementary firing takes place in a low-oxygen gas stream, this form of combustion will often produce significantly lower levels of NO_x than boilers using ambient air.

Furthermore, the combined NO_x emissions generated by the prime mover and the supplementary firing unit will usually be much lower than those that would arise if both plant items were operated separately. Supplementary firing is usually carried out using in-duct burners, although conventional boiler burners may be used in conjunction with water-tube heat recovery boilers. In the case of reciprocating engine sets, the supplementary firing facilities must be designed to be able to operate satisfactorily with a pulsating exhaust gas flow.

Auxiliary firing involves the provision of an air supply to the supplementary burner in place of the turbine or engine exhaust gases, enabling the boiler to provide heat energy to the site when the cogeneration generator set is not operating. Thermal efficiency will be lower than it is for conventionally fired boilers, but this is of marginal significance as long as operation under these conditions represents only a small proportion of the total running time. The availability of auxiliary firing can avoid the need for other standby boiler plants.

Supplementary and auxiliary firing both improve the overall cost-effectiveness and the flexibility of cogeneration plant. However, there are restrictions on the extent to which this approach can be used, and these restrictions are determined by limitations that are imposed by the materials or by the construction of the heat recovery boiler. Supplementary and auxiliary firing entails additional capital costs, and this, in conjunction with the operating cost savings, has to be compared with the alternative of maintaining conventional boiler plant for heat top-up or standby purposes.

4.5 Heat output from the cogeneration package

It is important to consider heat distribution in conjunction with heat recovery, as the distance between the heat load and the cogeneration unit influences the cost of the pipework that is needed to connect the cogeneration unit to the site. Unless the exhaust gases are used directly for heating or drying, the heat output from packaged cogeneration units is usually in the form of hot water. The heat is transferred to the user via a closed loop 'flow and return' pipework system.

The flow pipework delivers hot water at about 80 °C to the point at which the heat energy is transferred to the user. The water then passes into the return pipework, and is returned to the cogeneration package at a temperature that is about 10 °C lower than that of the flow temperature. The closed nature of the loop means that the hot water is not used directly, but instead acts as a heat transfer medium. This allows for the addition of small quantities of chemical to the water to improve the system's resistance to frost and corrosion.

In many cases, the cogeneration package will be installed on a site where a hot water flow and return system already exists, and it may be possible to make the necessary connections to the appropriate parts of the existing circuit. There are essentially two ways of connecting a cogeneration unit in this situation:

1. in series, as a bypass in a suitable return to the boilers;
2. in parallel with the boilers.

Connection in series is most frequently used with existing installations, since it creates minimum interference with the existing flow and control arrangements. Connection in parallel is preferred for completely new installations, especially where the cogeneration unit is likely to supply a significant proportion of the total heat load. In both cases, it is usually possible to connect the cogeneration package into the existing heat system in such a way that it adds its heat upstream of the existing boilers or water heaters. The existing boilers then operate as top-up or standby facilities for the cogeneration plant.

When the heat output from the cogeneration plant cannot be used on the site and the power output must be maintained, a cooling system needs to be incorporated within the flow and return pipework. This is often referred to as a 'dump radiator', and it is normally controlled by a valve that is connected to a temperature sensor on the return water inlet to the cogeneration package. If the water temperature exceeds a set level, then the valve opens to pass water into the dump radiator and directly back to the return pipework system.

In order to comply with the requirements for good quality cogeneration, a suitable method for measuring the heat energy supplied to the site (rather than dumped) must be provided. The hot water system must be designed to achieve the rates of flow and the return water temperature that will allow continuous operation of the cogeneration package. The system pipework must be of the correct diameter, and it must incorporate sufficient pumping capacity to maintain the correct flow and temperature conditions. It is common to equip the system with duty and standby pumps to ensure maximum availability.

Furthermore, the pumps must be selected to operate with the dump radiator system in either full or partial use, or with the hot water flow all passing to the site. It is also important to ensure that heat distribution systems have sufficient levels of thermal insulation to prevent heat loss and minimise hazards. The system must also incorporate the means of isolating individual plant items for maintenance, while allowing others to continue to operate.

4.6 Utility interconnections

A cogeneration plant is usually connected to the site electrical system in such a way that it can operate in conjunction with the local area electricity supply system. This is achieved by closing the electrical switchgear connections between the cogeneration plant, the site and the local area system, with the cogeneration plant and the local area system operating electrically locked together. This is known as a parallel mode operation of the system.

There are also a number of cogeneration plants that are installed without an electrical connection to an external electricity system, often arising as a necessity as a result of the site's location or special circumstances. These sites are said to operate in island mode. They have the added benefit of avoiding the costs of installing external site connections, but they have to manage their provision and consumption of power with no top-up or back-up supplies. This usually requires a high level of installed plant capacity in order to ensure that there is power availability at all times. However, many of the sites that operate cogeneration plants in parallel mode also have the facility to operate in island mode.

This provides them with the particularly useful capability of being able to provide power to the site when the local area electricity system has suffered a supply failure. In normal parallel mode, all of the circuit breakers are closed. If the local system fails, the cogeneration plant feeds the site load, which can be limited in order to match the capacity of the cogeneration plant. When the local system is restored, the cogeneration plant is synchronised with the local system.

When a change from parallel to island mode occurs instantaneously, it should be possible for the cogeneration plant to continue to supply the site load without any interruption, provided that the site load can be immediately limited to the output level of the cogeneration plant. This is usually achieved by using load monitoring and control equipment, which can automatically disconnect selected parts of the site load. If this load limitation cannot be achieved, then the cogeneration plant will usually shut down when there is a failure in the local area system with which it is operating in parallel. The site will then lose all power supplies.

However, as long as the site system has facilities to disconnect selected supply circuits within the site, the cogeneration plant can be quickly restarted to provide site power up to its maximum output level. This usually requires prompt action by the site staff in operating the circuit breakers according to a prepared procedure. It is also necessary for the cogeneration plant to be equipped with a back-up power source. The back-up power source will usually comprise a small standby diesel generator. If the site has other standby generation facilities, then power can be made available from these sources to restart the cogeneration plant.

The requirements for interconnection with public utilities' grids vary from country to country, from operator to operator, and from utility to utility, depending on the generation equipment, size and host utility systems. Interconnection equipment requirements increase with the generator size and voltage. In general, the complexity of the utility interface depends on the mode of transition between parallel and stand-alone operation. The plant that is connected to the electric grid must have an automatic control utility tiebreaker and the associated protective relays.

When a cogeneration system is integrated into the utility system, a number of issues have to be taken into account, such as control and monitoring, metering, protection, stability, voltage, frequency, synchronisation and reactive compensation of power factor, safety, power system imbalance, voltage flicker and harmonics. Providing utility interconnection with on-site cogeneration technologies is currently a lengthy process.

4.7   Future market development

There are several factors that will affect the growth of cogeneration activities. These factors include the initial cost of buying and bringing a cogeneration system online, the maintenance costs and the environmental control requirements. Some utilities do not have a need to purchase extra electricity, which can reduce the viability of cogeneration projects that rely heavily on power sales to utilities.

Cogeneration facilities are most likely to be economically viable at locations where there is a coincident demand for electricity and thermal energy during most of the year and that also have easy access to cheap fuels. Typical markets for cogeneration include:

- energy-intensive industries, such as the chemical, refining, forestry products, food and pharmaceutical sectors;
- district energy system that distributes heat or hot or cold water to a network of buildings;
- high-power reliability/quality applications, such as internet or telecommunications data centres that require high quality, reliable power and substantial cooling capacity;
- institutional markets, such as hospitals, hotels, conference centres and leisure centres;
- abandoned industrial sites, where cogeneration plants can provide the energy infrastructure that is required for supporting power plants and facilitating economic redevelopment of underutilised properties;
- domestic cogeneration market, which offers major potential opportunities as domestic cogeneration technologies become more cost-effective.

It is in these areas that the most significant technical developments and market growth are likely to take place.

4.8   Government support

In the UK's March 2004 budget, the UK government reduced the VAT rate on domestic micro-cogeneration appliances to the same level as other energy-saving measures. This came into effect in 2005. This move was designed to encourage people to invest in this technology. The UK's 2006 budget made an extra £50 million available to micro-cogeneration under the new Low Carbon Building Programme, in addition to the £30 million over 3 years that had been made available in November 2005, so that local authorities, schools and other public bodies made increased use of micro-cogeneration, thus providing an acceleration to savings in the cost of producing such units. David Sowden, the chief executive of the Micropower Council, said:

> This is a real boost to the micro-cogeneration industry. We have long been putting the case for market transforming policies of this nature. We are very pleased that the government is taking such a strong interest in the scope for micro-cogeneration to reduce carbon, save consumers energy, and change attitudes in the way people regard their own use of energy.

It has been estimated that the use of micro-cogeneration could cut a typical household's energy bills by up to £150 a year and CO_2 emissions by up to 1.5 tonnes a year. In early 2003, the UK government, in its Energy White Paper, identified energy efficiency as being the cheapest, cleanest and safest way of meeting its Kyoto and manifesto targets.

The German Parliament has adopted a new Law on Renewable Energy, which includes the provision of incentives to encourage significant improvements for cogeneration installations. The law will provide particular incentives for biomass-fuelled cogeneration plants, and it will give all operators of cogeneration plants a more reliable basis on which they can calculate the value of electricity that is being exported to the grid. The new law specifies that installations that burn biomass to produce electricity will obtain an additional 2 Eurocents per kilowatt hour of electricity produced if they operate in cogeneration mode. In addition, certain other technologies are eligible for an additional 2 Eurocents per kilowatt hour.

4.9 Using cogeneration in hospitals

One of the largest users of energy in the UK is the health sector. There are approximately 1 200 NHS hospitals in the UK, and the cogeneration potential for these has been estimated to be greater than 1 000 MWe. There are many reasons why the use of cogeneration is a good idea in a hospital. It can be one of the most effective ways for the hospital to save money, by reducing energy costs. It also ensures greater reliability of supply of energy. In addition, as cogeneration reduces energy waste, it can help to reduce health and environmental problems arising from emission of pollutants. Finally, cogeneration can be used to dispose of medical waste that would otherwise need to be disposed of at some expense.

4.9.1 Save money

The main benefit of cogeneration to a hospital is the financial saving that can be made. Cogeneration can reduce the energy bill significantly. In typical cases where a hospital has installed a cogeneration plant, the experience has been that savings of between £33 000 and £350 000 per year have been achieved.

However, these figures must be offset against the cost of installing a cogeneration plant. This varies according to the choice of payment method. As a general rule, cogeneration systems regularly achieve a 3–5 year payback, and over their lifetime, internal rates of return above 15 per cent can be achieved.

4.9.2 Improved reliability

It is inevitable that breakdowns and maintenance will affect heating and power supply. By augmenting or replacing the existing plant, cogeneration can considerably improve the quality and reliability of the energy services that are being provided. Cogeneration can also be configured to increase the capacity of on-site standby generation in the event of a power cut.

4.9.3 Finance

If the hospital buys the cogeneration unit itself, it may be limited to a smaller unit, depending on the available resources. Energy services companies can provide the necessary capital for cogeneration, and for additional work if necessary. As the hospital pays only for the energy that it requires, all responsibility for checking the economic feasibility and ensuring savings lies with the contractor.

There are three main ways of funding a cogeneration plant at a hospital:

1. capital funding;
2. energy services approach, using either discount energy purchase or CEM;
3. private finance initiative (PFI).

The main benefit of capital funding is that complete control over the cogeneration plant can be retained, and the cost savings can thus be maximised. However, the hospital trust will retain the risk of the asset not performing.

Maintenance of the cogeneration plant can be contracted out, and this is usually to the equipment supplier, with a typical guaranteed level of availability over a 1–10 year period. Although it is slightly more expensive, this approach ensures that the expected savings are achieved and it minimises the trust's exposure to breakdown repair costs.

Capital funding should not be considered where the trust is not in a position to devote time and resources to the design, project management and ongoing audit of the installation.

The cogeneration study should consider the potential effect of capital charges on the project, including any avoided charges from plant replaced and reduced capital spend.

Alternative methods of funding of cogeneration in hospitals are becoming increasingly common. These enable energy and cost savings to be obtained without risking the capital. However, all of the options need careful scrutiny to ensure that there is appropriate risk transfer and value for money.

Although design costs and risks are removed under external funding schemes, the hospital must still ensure that it uses suitable expertise to scrutinise the technical, financial and legal proposals. Energy services contracts take the concept further by removing the hospital's responsibility for all energy services up to the point of use.

In entering into an agreement with an energy services company, the contractor maintains agreed environmental conditions in the building. This may be, for example, 21 °C and five air changes per hour. The means by which contract conditions are met is solely the responsibility of the contractor.

Discount energy purchase is also widely known as equipment supplier finance. In this option, the cogeneration plant supplier pays for, maintains and owns the installed cogeneration plant. The hospital then buys units of electricity from the cogeneration plant at a price that is less than that of mains electricity, but is higher than the cost of generating the electricity. This allows the supplier to recover its costs. Gas costs are usually paid for by the hospital, but the heat is normally provided free of charge.

A major benefit of discount energy purchase is that it guarantees a specified level of cogeneration operation and performance. Even if the cogeneration plant fails to perform, the hospital still makes savings, as the supplier must supply all the specified electricity demand at the contract price. These contracts normally run for five years and upwards, during which time the hospital buys a set amount of electricity. When the contract ends, the hospital either owns the cogeneration plant or can exercise a purchase option.

CEM is the system that is typically used for large, capital-intensive projects, including those where building work, refurbishment and improvements are required. A contractor designs and finances the work, and is contracted to provide the energy supply to all or part of the site. Installation of the cogeneration plant often forms part of the contract as it provides a controllable cost. The hospital buys heat and power from the contractor at a set tariff, which is normally in the form of a standing charge and unit rate for each service. A profit margin on the unit price of this energy and the standing charge allow the contractor to recoup the cost of the investment. At the same time, the hospital saves money via reduced overall energy costs. Contracts tend to be in excess of 7 years in order to reduce the capital element of the costs. Once the contract comes to an end, the plant can then become the property of the hospital, subject to accounting treatment.

Funding of a cogeneration supply in the UK may be possible under the PFI. The principal advantage of PFI is that the operational risks are transferred to the private sector in return for a regular service charge.

A selection procedure to determine the best payment option for a cogeneration plant is shown in Figure 4.1.

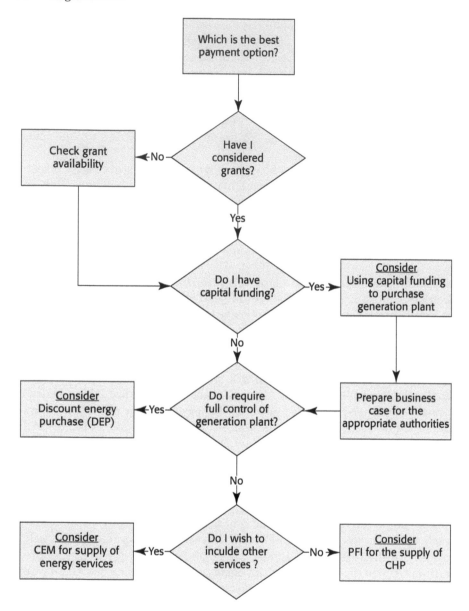

Figure 4.1   Selection of payment option

Chapter 5
Can we use cogeneration?

As discussed in Chapter 4, many different types of organisation and even households might find cogeneration to be a practical and viable method of energy supply. While these different organisations may have different priorities and requirements, there is a basic process by which they can determine whether a cogeneration system would be viable and economically competitive.

The best way to assess the attractiveness of a cogeneration project is to conduct a detailed financial analysis and compare the returns with the market rate for investments in projects presenting similar risks. Well-conceived cogeneration facilities should incorporate technical and economic features that can be optimised to meet both heat and power demands of a specific site. A comprehensive knowledge of the various energy requirements of the site, as well as of the cogeneration plant characteristics, is essential in order to derive an optimal solution.

As a first step, the compatibility of any existing thermal system with the proposed cogeneration facility should be determined. This will help the designer make use of the existing infrastructure. Important user characteristics that need to be considered include electrical and thermal energy demand profiles, the prevalent costs of conventional utilities and any physical constraints of the site. A factor that should not be overlooked at this stage is the need for reliable energy supply, as some industrial processes and commercial sites are very sensitive to any disruption of energy supply that may lead to production losses.

To fully exploit the cogeneration facility throughout the year, potential candidates for cogeneration technology systems should have the following characteristics:

- adequate thermal energy needs, matching with the electricity demand;
- reasonably high electrical load factor and/or operating hours;
- fairly constant and matching electrical and thermal energy demand profiles.

These are essential for the full exploitation of the cogeneration installation. In addition, part-load operation of the plant should be avoided where possible, because this would negatively impact the economic viability of the project.

A cogeneration project is the same as any other commercial energy efficiency project that requires high investment, has a relatively long payback period and presents some potential financial risks. Therefore, the steps that would normally be followed in developing a cogeneration facility would be the same as those that are employed for any investment project (see Figure 5.1). Projects will obviously vary from one to another on the basis of who is the project developer, what is the size of the project, who is financing the project and so on.

Prior to undertaking any economic analysis to assist in determining the potential commercial benefit of a cogeneration project, there are a number of parameters that have to be considered:

- heat-to-power ratio
- the quality of thermal energy needed
- electrical and thermal energy demand patterns
- fuel availability
- required system reliability
- local environmental regulations
- dependency on the local power grid
- option for exporting excess electricity to the grid or to a third party

Cogeneration represents a major capital investment, and thus it is always in competition with other energy efficiency projects. Organisations usually have limited resources to devote to energy efficiency projects, and cogeneration is only one such approach. Although the principle of cogeneration is relatively straightforward, the development process for any given cogeneration installation consists of a number of separate steps. Each one of these steps is an essential component of the whole development procedure and must be competently and thoroughly undertaken if the project is to be successfully designed and installed, and then operated effectively throughout its lifespan.

The development process requires a significant commitment in terms of both time and expertise, thereby incurring both cost and effort. The required expertise is special and covers several disciplines. Proper management of the process is important if the proposed cogeneration plant is to maximise the potential benefits to its future operator. The instigator and manager of the development procedure will need to ensure that the necessary managerial skills are made available and are applied effectively.

The following should be borne in mind when considering a cogeneration scheme:

- Adequate preparation is the key to a successful installation. The arguments in favour of cogeneration are complex to both engineers and non-engineers. Therefore, any project proposals need to be rational and clear and need to consider all the arguments and options.
- The financial returns have to be at least equal to other potential capital projects.

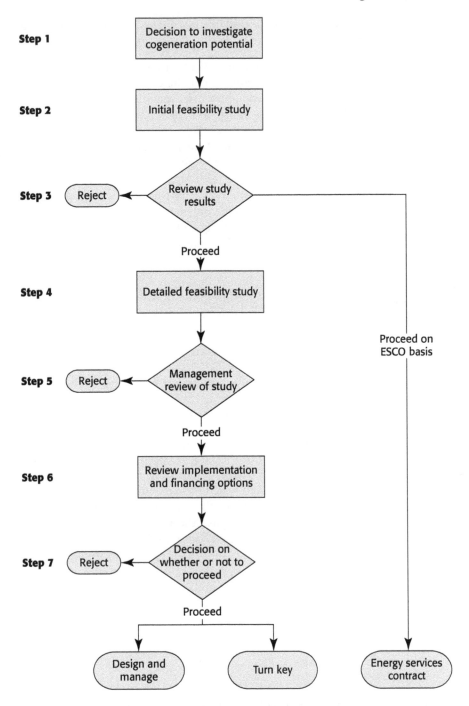

Figure 5.1   Typical steps for cogeneration project development

- There are a number of different available methods of financing and operating cogeneration, each of which has its own distinct benefits and drawbacks.

There is one basic question that every potential cogeneration operator faces: whether it is viable and profitable under the circumstances that pertain to that particular operator. The secondary question that potential operators need to answer is whether any of the alternative methods of obtaining energy is more cost-effective than cogeneration.

There are several factors that are involved in determining the suitability and viability of a cogeneration plant for any given application. These can be summarised in the form of a checklist, although not every element of the checklist is required to make a cogeneration viable for a particular application. As a rough rule of thumb, the more items on such a checklist that apply, the more profitable a cogeneration plant is likely to be.

Factors involved in determining the viability of a cogeneration plant include the following:

- fuel availability;
- demand for heat, steam and electricity;
- need for a secure supply;
- need to reduce emissions levels;
- a desire to decentralise power supply from the national grid;
- transport issues;
- the desire to be able to both receive and export electricity to the national grid;
- need to dispose of locally produced by-products.

Many types of potential operators might well be in a position to benefit from the use of cogeneration, and they should give careful consideration to this option. As an example, industrial processes that produce burnable by-products that would need to be disposed of, and which have a demand for electricity and heat or steam, are likely to benefit from the use of cogeneration. It makes logical economic sense for such users to use the waste that they generate to create profit rather than having to pay to dispose it of.

One of the big advantages offered by cogeneration is that it uses fuel very efficiently. This makes it especially appropriate for use on sites where it is important to minimise the amount of fuel used to produce a given amount of energy. This could be for a variety of reasons:

- high fuel cost
- unreliable fuel supply
- fuel transport difficulties
- variable fuel costs

Many governments – especially in Europe – are keen to encourage the development of cogeneration as it is one of the most effective means of reducing emissions levels and achieve the reduction targets that have been set. As a result, there are a number of financial incentives that are available from many governments that are intended to encourage cogeneration development. According to the UK's Energy Savings Trust, in 2003, about 96 per cent of funding for cogeneration schemes at hospitals and universities in the UK came from banks (72 per cent) or from community energy schemes (24 per cent). The British government is keen for Private Finance Initiatives (PFIs) to provide finance for such schemes. To date, a few hospital PFI schemes have included cogeneration plants in their scope, but the Energy Savings Trust believes that there is significant potential in this area.

It also believes that cogeneration systems can deliver cost savings on a whole-life basis, along with environmental benefits, to community heating systems. There are a number of examples of cogeneration that are being used effectively at hospitals:

- Down Lisburn HSS Trust, Northern Ireland
- Diana, Princess of Wales Hospital, Grimsby
- Southampton University Hospital
- Leeds Teaching Hospitals

5.1 Operation modes of cogeneration systems

A mode of operation is determined by the criterion on which the adjustment of the electrical and useful thermal output of a cogeneration system is based. There are various modes of operation possible:

- *Heat-match mode*: The useful thermal output of a cogeneration system at any instant in time is made equal to the thermal load, without exceeding the capacity of the cogeneration system, by adjusting the output level to suit the heat demand. If the electricity generated as a result is higher than the load, then the excess electricity is sold to the grid; if it is lower, supplementary electricity is purchased from the grid.
- *Base thermal load matching mode*: The cogeneration system in this mode is sized to supply the minimum thermal energy requirement of the site, and it operates at all times. Standby boilers or burners are operated to satisfy heat demand above the minimum base level. If the electricity demand of the site exceeds that which can be provided by the prime mover, then the remaining amount can be purchased from the grid. Likewise, if the local laws permit, any excess electricity can be sold to the power utility.
- *Electricity-match mode*: The generated electricity at any instant is equal to the electrical load, without exceeding the capacity of the cogeneration system. If the cogenerated heat is lower than the thermal load demand, then an auxiliary boiler is used to supplement for the needs. If it is higher, then excess heat is rejected to the environment through the coolers or the exhaust gases.

- *Base electrical load matching mode*: In this configuration, the cogeneration plant is sized to meet the minimum electricity demand of the site on the basis of the historical electricity demand curve. The rest of the needed power is purchased from the grid. The thermal energy requirements of the site could be met by the cogeneration system alone or by additional boilers. If the thermal energy that is generated with the base electrical load exceeds the plant's demands, and if the situation permits, then any excess thermal energy can be exported to neighbouring customers.
- *Mixed-match mode*: In certain time periods, the heat-match mode is followed, while in the other the electricity-match mode is followed. The decision is based on various considerations such as the load levels, the fuel price and the electricity tariff at the particular day and time.
- *Stand-alone mode*: There is complete coverage of the electrical and thermal loads at any instant with no connection to the grid. This mode requires the system to have both reserve electrical and thermal capacity, so that in case one unit is out of service for any reason, the remaining units are capable of covering the electrical and thermal load. This is the most expensive strategy.

In general, the heat-match mode results in the best fuel energy saving ratio, and possibly has the best economic performance for cogeneration in the industrial and building sectors. In the utility sector, the mode of operation depends on the total network load, the availability of power plants and the commitments of the utility with its customers regarding the supply of electricity and heat.

However, applying general rules is not the most prudent approach in cogeneration. Every application has its own distinct characteristics. Cogeneration systems can come in a variety of designs, depending on the type of technology, size and configuration used. The design of a cogeneration system can be tailored to the needs of the user, including the possibility of changing the modes of operation. Moreover, the technical and economic parameters may change with the day and time during the operation of the system.

All of these aspects mean that it is necessary to optimise cogeneration systems through the use of microprocessor control systems. These provide the capability to operate the cogeneration plant in base-load mode, track either electrical or thermal loads or operate in an economic dispatch mode.

In the latter mode, the microprocessor can be used to monitor cogeneration system performance, including:

- system efficiency and the amount of useful heat available;
- the electrical and thermal requirements of the user, the amount of excess electricity which has to be exported to the grid and the amount of heat that must be rejected to the environment;
- the cost of purchased electricity and the value of electricity sales, as these can vary with the time of day, the day of the week or the season.

On the basis of this data, the microprocessor can determine which is the most economical operating mode, or even whether the unit should be shut down. In addition, by monitoring various operational parameters such as efficiency, operating hours, exhaust gas temperature and coolant water temperatures, the microprocessor can help in maintenance scheduling. If the system is unattended, a telephone line can link the microprocessor with a remote monitoring centre, where the computer analysis of the data may notify the skilled staff about an impending need for any scheduled or unscheduled maintenance. Furthermore, as part of a data acquisition system, the microprocessor can produce reports of system technical and economic performance.

5.2 Distribution of heat

The heat generated from the cogeneration system can be distributed under different conditions. The best thermal distribution for the HVAC system must be suitably selected and its cost and dimension must be well designed. The various thermal distribution systems are summarised in Table 5.1.

Table 5.1   Thermal distribution systems

System	Applications	Advantages/disadvantages
Steam		
Low pressure	Heating system	Lower heat loss than high pressure system
	Single-stage absorption chiller	More complex start-up and shutdown
High pressure (1 000–1 400 kPa)	Two-stage absorption chiller	High temperature available
		With condensate return: high construction, low operating cost
		Without condensate return: low construction, high operating cost due to the loss of condensate thermal energy and compensate water and its chemical treatment
		More complex start-up and shutdown
Hot water		
High temperature (up to 200 °C)	Heating system	Smaller pipes than high-pressure system
	Single-stage absorption chiller	Lower operating costs
	Large HVAC systems	
Low temperature (less than 80 °C)	Heating system	Lower heat loss
	Single-stage absorption chiller	
	Large HVAC systems	

It is important to consider heat distribution in conjunction with heat recovery, as the distance that lies between the heat user and the cogeneration plant, and the form in which the user requires the heat to be delivered (steam, hot water, space heating), will influence the design of the cogeneration plant. In the case of direct use of the heat from exhaust gases, the ductwork for transporting the exhaust gases is relatively bulky and must be well insulated. Installation costs will limit the distance over which a system of this type can be used.

Furthermore, the performance and use of gas turbines and, to a lesser extent, of reciprocating engines are adversely affected by the exhaust system back-pressure, and this also limits the length of exhaust ductwork that can supply direct heat. The same limitations apply to the siting of an exhaust heat recovery boiler, and it is normal for a cogeneration plant to be located adjacent to an existing central boiler plant, and is often integrated with the boiler plant.

The heat that is recovered from a cogeneration plant is frequently distributed using the existing systems connected to the central boiler house, and the existence of an existing system is an important factor in determining the cogeneration plant design and location. On decentralised sites, where heat uses are widely dispersed, the size and location of the cogeneration plant are constrained by the available heat demand within the adjacent area.

It may be cost-effective to recover heat from a cogeneration plant and distribute it in a single, thermally efficient form such as steam or thermal oil. However, there is a limit to the distance that such systems can cover without the costs of installation and operation becoming prohibitive. Therefore, on many decentralised sites, it may be more beneficial to install several cogeneration units adjacent to the users of the recovered heat, but this does require a fuel distribution system and also requires the availability of suitable connections to site electrical systems.

5.3   Silencers

Many cogeneration systems, because of the prime mover selected, generate noise and vibration. This can cause an environmental problem, especially where cogeneration plants are located close to domestic dwellings or in non-industrial settings. Where environmental issues arise as a result of the noise generated, some means of attenuating the noise has to be found. There are many designs in the market for all kinds of plants ranging from simple exhaust silencers for small reciprocating engines to large, integrated complex designs for multi-megawatt power plants.

Acoustic housings and enclosures range in design from single-skinned acoustically lined panels to multi-inch thick composite wall designs that provide about 40–60 dBA reductions. Acoustically treated containers can be furnished with inlet/outlet silencers, specially treated acoustic walls, air-handling equipment, fire suppression, special lighting, extensive exhaust-silencing systems, fuel sub-bases, acoustically treated inlets/outlets, special hardware and any other option that the plant may require.

Other methods of noise attenuation include requiring that all piping to and from the engine be supplied with flexible connectors. Typical market products to meet these needs are stainless-steel exhaust connectors that incorporate engine-mating and silencer-mating flanges, 'Y' exhaust connectors for dual-outlet engines that are cost-effective and highly adaptable, and stainless-steel braided connectors for flexibility in rigid water and fuel lines that suit any flange or piping arrangement.

Vibration isolators are designed specifically for rugged generator set applications, and they provide an inexpensive way of isolating engine and genset vibration from buildings and structures. Slide dampers, level dampers and a non-skid sound pad are the usual features of such designs. Flexible connectors provide little compensation for thermal expansion in exhaust piping. For example, an 8-inch diameter exhaust pipe will expand by approximately 7 inch per 30 m at a temperature of 450 °C. To compensate for this expansion, multiple-ply stainless-steel expansion bellows are used to allow for axial expansion. Moreover, when an exhaust pipe or tube exits a building, the pipe should be isolated from the wall. For this purpose, wall and roof thimbles are used with features such as self-supporting roof thimbles incorporating the exhaust silencer and roof-mounting into the design of the thimble.

5.4   Factors to consider

The checklist shows the many factors that need to be considered to determine the viability of a cogeneration plant.

5.4.1   Fuel supply

- Does the site have easy access to cheap or free fuel supplies, such as waste from an industrial process that is carried out at the site?
- Does the site need to dispose of waste that can safely be burnt to reduce volume?
- Is the price or availability of fuel likely to rise or fluctuate sharply?
- Are there concerns about a particular fuel supply being able to meet a potential increase in demand?
- Are there any limitations imposed on fuel supply through transportation bottlenecks?
- How reliable is the generation of power? This mainly applies to weather-based renewable generating systems.

5.4.2   Demand

- Does the site have to generate both electricity and heat as part of its normal operation?
- Does the site have a demand for process steam and electricity?
- Is the cost of buying electricity from the grid excessive?

- Is there an asynchronicity in the demand from the site and from the grid, leading to the situation where it is cost-effective to buy electricity from the grid during off-peak grid demand periods and selling electricity to the grid at peak demand periods? Would a cogeneration unit capable of meeting some but not all of the peak demand at the site be the most cost-effective solution?
- Is there a local demand for heat, steam, high-purity water or electricity local to the site that could be met by a cogeneration unit?
- Is demand for heat and electricity constant, or does it vary significantly?
- Is the proportion of heat/electricity demand constant or fluctuating?
- Is there a desire to secure self-sufficiency in energy supplies?
- How acceptable – or otherwise – are sudden outages?
- Are the heat or power demands of the site likely to grow significantly? Can a cogeneration plant be expanded to meet increased demand?

5.4.3   Regulatory

- Might the site be required to make significant cuts in carbon emissions?
- Is this site likely to have to undertake measures to comply with pollution control regulations?
- Are there economic benefits to be gained from the sale of carbon credits?
- Is the central government keen to decentralise power? If so, does it make it easier to obtain planning permission for cogeneration plants?
- Is it easy to acquire all the necessary consents and permissions?
- Is the local authority keen to develop cogeneration, and is it willing to provide assistance?

5.4.4   Subsidies and financing

- Are there government subsidies available to encourage conversion to high-efficiency energy production?
- Can a cooperative be formed to help spread the cost of installation and operation of a cogeneration plant?
- Are carbon credits available, and how much revenue would their sale generate?
- Is the central government likely to impose significant carbon taxes that cogeneration would avoid?
- Is the local electricity utility in need of extra supply?
- Does the local utility buy power?
- How expensive is electricity bought in from the local utility?
- Are there grants for promoting green energy available? These could come from local authorities, central government, non-government organisations or regional authorities (such as the EU).
- Are there systems currently in place that need upgrading?

5.4.5 *Environmental*

- Is the central government likely to impose stricter environmental restrictions on power generation, and if so, are cogeneration plants likely to meet these restrictions?
- Is the organisation keen to apply sound environmental policies?
- Are there specific local environmental issues that may impact construction of a cogeneration plant?
- Is space at a premium at the site?
- Does access make transport of large amounts of fuel a problem?
- Is significant growth expected at the site?
- Are there likely to be local objections to construction?
- Is waste disposal an issue? Is cogeneration likely to reduce or eliminate the need for waste disposal?

5.4.6 *Construction and operation*

- How quickly can a cogeneration plant be brought on line? Does this meet with the demands of the site?
- Is construction likely to lead to significant temporary local employment?
- Are the necessary skills required for construction available?
- Are the necessary skills to operate the plant available?
- What is the expected life of the plant?
- Is plant decommissioning likely to be an issue?

5.5 What systems are suitable?

The two main types of cogeneration technique available are topping cycle plants and bottoming cycle plants (as described in Chapter 2). Topping cycle plants concentrate on generating electricity and sell any excess electricity to an electricity utility. These plants are typically sized to meet the heat demands of a site, with considerations of electricity generation being secondary. These topping cogeneration plants always involve the use of additional fuel, so there is an operating cost associated with the power production.

Bottoming cycle plants are much less common, and they are mainly restricted to heavy industries that use very high furnace temperatures and produce considerable amounts of waste heat. A waste heat recovery boiler can be used to recapture some of this waste heat and put it to productive use by producing steam that can be used to generate electricity through a steam turbine. No additional fuel is required to generate this electricity, as the fuel is required in the manufacturing process in any case.

Cogeneration systems are especially competitive when a site has to replace or upgrade a boiler unit. This reduces the cost of installation significantly, as the work needs to be carried out regardless of the type of heat source installed.

Cogeneration systems become even more cost-effective when it is easy to both buy and sell electricity from the local utility, as is often the case in liberalised energy markets. This requires a means to measure electricity flow both to and from the utility, which in turn implies further development of smart two-way metering systems. Such metering systems are being developed, especially for the domestic market, but at present, few have reached commercial status. Should they do so, then domestic-sized cogeneration systems will become more commercially attractive. At present, however, domestic cogeneration is not yet viable on a commercial scale although there are a number of field trials being undertaken in this field, particularly in Germany.

5.6   Project risks

There are five main categories of risk associated with a cogeneration project:

1. construction risk
2. operation risk
3. fuel supply risk
4. off-take risk
5. political risk

Construction risks include:

*   time, cost, quality
*   sponsor's ability to arrange construction
*   experience and standing of contractor
*   soundness of the technology
*   terms of the main construction contract (fixed price/turnkey; penalties/ bonuses; incentives; force majeure; commissioning; sub-contracting)

Operation risks include:

*   experience of the operator
*   maintenance of performance levels
*   cost constraints

Fuel supply risks include:

*   availability
*   specification
*   price

Off-take risks include:

*   quantity
*   capacity charge/energy charge
*   base unit price
*   indexation
*   term of contract

- take-or-pay
- penalties for non-supply
- status of off-taker

Political risks include:

- legislation
- taxation
- environmental controls

The identification, quantification and allocation of these risks are a specialised skill requiring technical, financial and insurance advice. The contractual approach to project implementation should enable the risks to be correctly identified and allocated. It is essential that the risks should be allocated to the party that is best able to control those risks. A number of options are available to the owner when implementing the cogeneration scheme. For example, if the owners have ample resources (which is an unlikely scenario), they may choose to manage the whole project.

A more likely approach is for the owners to appoint a consulting engineer to do this on their behalf, which would then define the scope of the project, specify its performance and monitor the progress of construction and financial expenditure. The owner, either by himself or with the assistance of a consulting engineer, may decide to place a turnkey contract for the whole scheme. In this case, the consulting engineer would audit the progress of the contract and give the owner reassurance that the work will be completed on time and to specification.

It is clear that the project manager will have a major responsibility not only for the management of the technical interfaces involved in the project, but also for the various complex commercial interfaces between the various contracts (real or pseudo) making up the entire scheme. There is no single implementation methodology that is correct for a particular size of scheme or for a particular industry application.

There are many schemes that have been examined from a technical standpoint and that have been shown to give good internal rates of return, but have failed to obtain corporate consent. In most of these instances, failure could have been avoided if the scheme had been delineated as a series of interlocking contracts. A scheme can be successfully implemented only with careful commercial analysis and the apportioning of risk across the interfaces between the contracts.

Potential owners of such schemes, unless they are already operating as successful energy supply companies, should seek professional advice at the outset. The advice should cover not only the technical aspects but also the financial aspects, risk identification and allocation and the possible sources of fuel. Cogeneration schemes on large sites can bring considerable financial benefits, but experts in the field should prepare the case for their adoption and the planning of their implementation.

Chapter 6
How do we implement cogeneration?

Prior to any decision to proceed with a cogeneration plant, the potential changes in the site energy requirements must be thoroughly investigated. Energy-saving measures, demand-side management procedures and any changes in processes can not only be cost-effective, but may also affect the type, size and economics of the cogeneration system. The selection of the optimum cogeneration system should be based on criteria that are specified by the investor and the user of the system, considering economic performance, energy efficiency, uninterrupted operation or other performance measures.

This can be summarised in a set of decisions that have to be made regarding:

- the type of cogeneration technology (steam turbine, gas turbine, reciprocating engine);
- the number of prime movers and the nominal power of each of them;
- heat recovery equipment;
- the need of thermal or electricity storage;
- interconnection with the grid (one-way, two-way or no connection at all);
- operation mode of the system.

Furthermore, the availability of heat may lead to the investigation of the feasibility of absorption cooling, which will affect the load and consequently the design of the system. Any decision should take into consideration the various legal and regulatory requirements that may impose limits on the design and operation parameters, such as the noise level, emission of pollutants and total operating efficiency. Developing and planning a cogeneration (or trigeneration, if absorption cooling is to be included) installation requires significant time, effort and investment. Hence, it is prudent to approach the task in a series of steps.

The whole process from initial conception to final design can be divided into three stages:

1. preliminary assessment
2. feasibility study and system selection
3. detailed design

The first stage requires less work, typically only 1–2 days, and it helps the designer to determine whether any further efforts are justified. The actions that need to be performed in each stage are described briefly in the following sections.

6.1   Preliminary assessment

An inspection of the site is carried out in order to reach a first approximation on whether the technical conditions are such that cogeneration could be economically viable. This stage is often called the 'walk-through analysis'. The aspects that are examined in this step are described further in the text.

6.1.1   Technical issues

- The level and duration of the electrical and thermal loads.
- The energy saving measures that could be implemented before cogeneration.
- Any plans for changes in processes that would affect electrical and/or thermal loads.
- The compatibility of the thermal loads with the heat provided by available cogeneration technologies.
- The effect that cogeneration may have on the need to install and operate other equipments such as boilers, absorption chillers or emergency generator.

6.1.2   Site conditions

- The availability of space for siting the cogeneration system.
- The ability to interconnect with the electrical and thermal system of the facility.
- Zoning and/or environmental limitations that would preclude cogeneration.

6.1.3   Economics

The average retail electric price, fuel costs and the required return of investment or payback should be carefully examined, in order to make sure that the fuel and electric rates support the development of a cogeneration plant.

With regard to the data collected during this stage, at the very minimum, the following data has to be collected, reviewed and analysed:

- *Electrical requirements*
  – Average demand during operating hours: _____kWe
- *Thermal requirements*
  – Form of thermal energy use: _____ steam _____ hot water _____ other

– What is the primary application for thermal energy at the plant?
_____

– Average demand during operating hours: _____
– Required conditions: _____
● *Operating conditions*
– Nominal operating hours per year: _____
– Number of hours per year that electrical and thermal loads are simultaneously at or above average values: _____
● *Energy rates*
– Average retail electric rate: _____
– Fuel price: _____

After finishing collecting and analysing all the data describing the plant's energy use, the designer should go through a logical progression about the site, as shown in Figure 6.1, which provides an introductory flow chart to help the designer evaluate the initial suitability of a cogeneration installation at a specific site. Once the progression of questions has been successfully completed, simplified cogeneration payback estimators can be used to determine whether the development procedure can proceed to the next step.

These steps refer to those that need to be considered in respect of an existing facility. However, similar aspects need to be examined when designing a new facility. In such a case, the integration of the cogeneration system with the rest of the installations is much easier and it has greater potential for improving the economic viability of the investment. A system can be designed to be more efficient when it is designed from scratch than when it has to interface with pre-existing conditions and limitations. In large projects, a pre-feasibility study might be advisable for a better assessment at this stage.

6.2 Feasibility study and system selection

This is the most crucial stage of the whole procedure, and it will determine whether cogeneration is viable and which is the best cogeneration system for the specific application. It includes the following actions:

● The collection of data and drawing load profiles for the various energy forms needed. Load profiles can be drawn for typical days of the week, for weekends, for various months and seasons.
● The collection of information about electricity and fuel tariffs, as well as about the legal and regulatory issues.
● The selection of the cogeneration technology that can provide the quality of heat required. The power-to-heat ratio may be an additional criterion for the selection, but not a very strict one, because it can be changed either by an additional equipment (augmented heat recovery, thermal storage, supplementary firing) or by a decision to cover part of the electrical or thermal load.

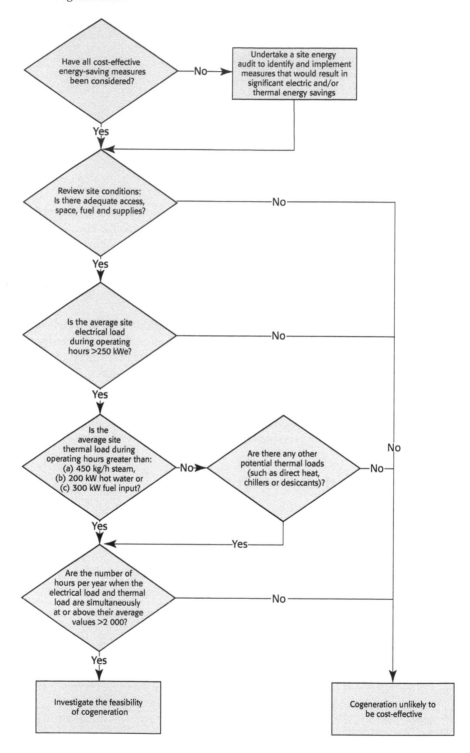

*Figure 6.1   Flowchart to evaluate the initial suitability of a cogeneration
installation at a specific site*

- The selection of the number of units and of the capacity of each unit. From the point of view of energy efficiency, the selection should be such that the major part of the cogenerated heat is used, thus avoiding its rejection to the environment.
- The selection of the operation mode and calculation of the energy and economic measures of performance. Calculations can be repeated for various operation modes.

The last three actions should be repeated for other combinations of technology, number and capacity of units, additional equipment and operation modes. At the end of this procedure, the system with the best performance is selected, and a single- or multi-criteria approach can be followed for this purpose. Moreover, a study of the environmental, social and other effects of the selected system has to be performed in this step of the procedure.

In addition to the data that was collected during the walk-through analysis stage, the following data must be collected, reviewed and analysed in this stage:

- *Electrical requirements*
  - Minimum demand during operating hours: _____kWe
  - Peak demand during operating hours: _____kWe
  - Annual electricity consumption: _____kWh
- *Thermal requirements*
  - Minimum demand during operating hours: _____
  - Peak demand during operating hours: _____
- *Energy rates*
  - Peak demand charge (if applicable): _____

In those cases where there is a strong phase shift between the electrical and thermal loads, the designer should examine the technical and economic feasibility of thermal storage or electrical storage to increase the utilisation of cogenerated electricity and heat. The selection of the type, the design and the control of the storage unit is of crucial importance for the energy and economic performance of the whole system.

After collecting and analysing the necessary energy and cost data, the next step is to select a potentially suitable cogeneration system. In doing this, a range of equipment data should be obtained from suitable suppliers, who will normally be willing to provide technical and cost information on their products. Attention has to be paid to the fact that the data that is related to the plant performance should reflect the typical running conditions of the plant over its expected lifespan, and not the peak operating conditions for the plant when it is new.

Information should be obtained for a number of options:

- Electrical output, which should include data relating to the power consumption of the cogeneration plant's own motors etc., so that the net output can be defined.

- Heat output that can be recovered for use on-site, including data on the flow rate and temperature of the fluid in which the heat is contained.
- Fuel consumption of the equipment, ensuring that this can be expressed in gross calorific value terms.
- The cost of supplying and installing the equipment.
- The dimensions and weight of the equipment.
- The approximate cost per kWh generated that should be allowed for servicing and maintaining the equipment.
- Any essential auxiliaries that are not contained within the scope of the equipment.
- After-sales services, such as on-site maintenance provision and availability of parts.

There are many variations of system structure and operation modes, and this makes an exhaustive search very difficult by conventional means. Several computer programs have been developed to aid the designer, and these are commercially available. These differ from each other with respect to the range of applicability and depth of analysis. One step further is the application of mathematical optimisation procedures for the system design and operation.

6.3   Identifying an appropriate output for a cogeneration plant

Initial selection of cogeneration plant is often dictated by two factors:

1. the site heat demand, in terms of quantity, type and temperature that can be met using heat from the cogeneration plant;
2. the base-load electrical demand of the site.

Sizing on heat demand will maximise the energy and environmental savings. Depending on the heat-to-power ratio of site energy demands, sizing to match the heat requirements will result in a scheme that may offer a surplus of electricity generation or it may require top-up electricity supplies during periods of peak demand. Under these conditions, the economics of exporting the electricity become a key issue in determining the economic cogeneration plant size.

Factors that have to be considered include the quantity of electricity that needs to be exported, the customer, the price that has been agreed, the time of day or year and any fixed up-front costs, such as upgrades that are required by the public electricity supplier and local electricity network. Once the electrical output of a cogeneration plant has been selected, the recoverable heat that is available from either a gas turbine or an engine can be compared with the site's heat demand. This comparison must take into account the temperature and form in which the heat can be recovered.

Ideally, this procedure will identify a size and type of the cogeneration plant where all of the electricity and recovered heat can be used on the site. In cases where all of the electrical power can be used but where there is a surplus of recoverable heat, it may be worth considering the use of a smaller unit. If all the recovered heat from the cogeneration plant can be used, but the electrical demand exceeds the output, then it may be better to select a larger unit. There must be sufficient total plant to meet the site heat demands at all times, and this sometimes results in the plant being sized to meet the site heat load, with the surplus electricity being available for export to the grid.

Ultimately, the selection of the prime mover and the electrical generator depends on the overall cost savings that can be achieved, and the aim of the calculation procedure is to identify the optimum plant selection. The unit value of avoided electricity purchase is greater than the value of recovering an equivalent amount of heat energy. Therefore, the selection of the cogeneration plant with the greatest energy efficiency is not always the one that provides the greatest financial benefits.

6.4 Detailed design

The next stage is a detailed study into the system selected. This detailed study requires a much greater degree of commitment. Its aim is to arrive at accurate information and results that will allow the project developer to make firm decisions about the technical, legal, commercial and financial viability of the proposed cogeneration scheme. Much of the work is similar, in principle, to that which was carried out in the initial feasibility study, and there should be no major change in the overall objective of the cogeneration evaluation procedure.

In order to produce a meaningful result, the level of detail and accuracy must be as high as possible. Thus, there may be a need to collect more accurate and detailed information about the load profiles and to repeat some of the actions of the initial feasibility study in greater depth. As a result of this more detailed study, some of the main characteristics of the system might end up being slightly modified. Detailed technical specifications of the main unit or units are written down, including capacity, efficiency and controls, as well as the emissions, noise and vibration levels. Specifications for other major components have also got to be prepared.

Finally, the location site of the system is selected, and the design study is performed producing the necessary drawings for the construction or modification of the building, if required, and for the foundation of the system. Construction drawings are also prepared for the fuel supply, including tanks if necessary, air inlet and exhaust gas ducts, piping, electric circuitry and grid interconnection. The study must be specified and prepared in enough detail to produce a final report and conclusions that will be sufficiently comprehensive for the managerial staff to decide whether or not to proceed with implementation of the cogeneration system.

6.5   Technical assessments

6.5.1   Site energy demands

It is important to be as accurate as possible in defining the future energy loads that need to be met by the cogeneration plant and by any other site energy supply plant. Past consumption data usually provides a good indication of future demands, but it is also important to take site-specific factors into account when assessing potential future energy requirements, such as potential site expansion. The energy demand data must be subdivided into a high level of detail – preferably down to daily demand profiles, which may vary according to the time of the year, day of the week and so on.

For heat use, the demand must be subdivided according to the different heat load temperatures and other demand conditions.

Specific data requirements include:

- *Electricity demand data*: The electricity supplier may be able to provide a load profile for a 24-h period, a week or even for a whole year. Where this is not possible, electrical current flows can be measured by using clamp-on meters that are linked to a data logger. Careful interpretation is necessary in order to account for the production and non-production periods, any variations that there might be in production trends, any seasonal changes in electricity demand, and so on, and to generate the appropriate annual, seasonal or daily profiles.
- *Heat demand data*: Heat loads can be more difficult to assess, particularly steam heat loads. It may be necessary to carry out a detailed survey of the various heat uses, including the type and grade of heat that is used in each case, and to combine this with hourly or half-hourly readings of boiler fuel or steam/heat meters over a typical 24-h period. The required heat load values can be derived from fuel flow rates and system loss calculations. It is important to remember that the measurements taken represent the situation at one specific time period. As is the case with electricity, careful interpretation is required to generate the appropriate annual, seasonal and daily load profiles.
- *Likely future changes in demand*: It is very important to assess the effect of future changes in demand, which are usually associated with the implementation of energy efficiency measures, the introduction of new production or other facilities, discontinued processes or changes in operation. As a result, heat loads are likely to be reduced by the future implementation of energy conservation measures or by the introduction of local metering, charge allocation and other incentives for plant operators. Consumption data forecasts are needed for the expected lifespan of the potential

cogeneration system, which will be a minimum of 10 years. If the demands are forecast to change, then the data for each year should be provided wherever possible. The greater the level of detail that is provided, the greater will be the potential accuracy of the subsequent financial evaluation.

6.5.2 Timing of demands

Since cogeneration produces both heat and electricity simultaneously, it is essential to consider the extent to which the site has concurrent heat and electricity demands that can use the outputs of a cogeneration installation. This requires a time-based assessment of the site's energy demands. Some sites have fairly constant levels of demand over long periods, with minor variations resulting from occasional changes in plant availability or site activity.

Other sites have defined and predictable changes in demand that are associated with regular working patterns. It is generally the case that sites that require large amounts of energy for space heating will usually show significant variations between their winter and summer demands. For the purposes of the initial feasibility study, it is sufficient to consider the site consumption over a one-year period, and then sub-divide this period into a maximum of eight time bands, according to the actual site demand conditions.

The split would typically be based on distinctions between:

- daytime and night time
- weekday and weekend
- summer and winter

Once a decision has been taken on the appropriate number of time periods to be used, the electricity and heat demand data must be assessed and recorded in such a way that they can be readily used to calculate the potential energy cost savings. While this will inevitably require a certain amount of averaging, this approach provides a sufficient degree of accuracy at this stage of the evaluation process.

6.6 Factors affecting selection of the suitable cogeneration plant

While selecting the appropriate cogeneration system for a specific site, a number of specific technical parameters should be borne in mind. The most important of these factors are as follows:

- heat-to-power ratio
- quality of thermal energy required
- fuel supply
- noise levels
- regulatory and local planning issues
- other issues

6.6.1 Heat-to-power ratio

Heat-to-power ratio is one of the most important technical parameters influencing the selection of the type of cogeneration system. The heat-to-power ratio of a facility should match with the characteristics of the cogeneration system that is to be installed. Heat-to-power ratio is defined as the ratio of thermal energy to electricity required by the consuming facility.

Basic heat-to-power ratios of different cogeneration systems are given in Table 6.1, along with some other technical parameters. As can be seen, the steam turbine cogeneration system can offer a large range of heat-to-power ratios.

6.6.2 Quality of thermal energy required

The thermal requirements of the end user may dictate the feasibility of a cogeneration system or the selection of the prime mover. Gas turbines offer the highest quality heat that is often used to generate power in a steam turbine. Gas turbines reject heat almost exclusively in their exhaust gas stream. The high temperature of this exhaust can be used to generate high-pressure steam or lower temperature applications such as low-pressure steam or hot water.

Very large gas turbines (typically above 25 MW) are frequently used in combined cycles where high-pressure steam is produced in the heat recovery steam generator, and this is then used in a steam turbine to generate additional electricity. The high levels of oxygen that are present in the exhaust stream allow for supplemental fuel addition to generate additional steam at high efficiency. Some of the developing fuel cell technologies, including molten carbonate fuel cells and solid oxide fuel cells, will also provide high-quality rejected heat that is comparable to a gas turbine.

Reciprocating engines and the commercially available phosphoric acid fuel cells produce a lower grade of rejected heat. Therefore, heating applications that require low-pressure steam or hot water are the most suitable for these technologies although the exhaust from a reciprocating engine can generate steam up to 7 bar. These engines typically have a higher efficiency than most

Table 6.1   Heat-to-power ratios and other parameters of cogeneration systems

Cogeneration system	Heat-to-power ratio (kWth/kWe)	Power output (% of fuel input)	Overall efficiency (%)
Back-pressure steam turbine	4.0–14.3	14–28	84–92
Extraction/condensing steam turbine	2.0–10.0	22–40	60–80
Gas turbine	1.3–2.0	24–35	70–85
Combined cycle	1.0–1.7	34–40	69–83
Reciprocating engine	1.1–2.5	33–53	75–85

gas turbines in the same output range, and they are a good fit where the thermal load is low relative to the electricity demand.

Reciprocating engines can produce low- and high-pressure steam from its exhaust gas although low-pressure steam or hot water is usually specified. Jacket water temperatures are typically limited to 100 °C so that jacket heat is usually recovered in the form of hot water. All of the jacket heat can be recovered if there is sufficient demand. However, only 40–60 per cent of the exhaust heat can be recovered to prevent condensation of corrosive exhaust products in the stack that will limit equipment life.

6.6.3   Fuel supply

A potential system issue for gas turbines is the supply pressure of the natural gas distribution system at the end user's property line. Gas turbines need minimum gas pressures of about 8 bar for small turbines, with substantially higher pressures required for larger turbines. Assuming that there is no high-pressure gas service, the local gas distribution company would have to construct a high-pressure gas line, or the end user must purchase a gas compressor. The economics of constructing a new line must consider the volume of gas sales over the life of the project.

Gas compressors may have some reliability problems, especially in the smaller size ranges. If 'black start' capability is required, then a reciprocating engine may be needed in order to turn the gas compressor, thus adding to the cost and complexity. Reciprocating engines and fuel cells are more accommodating to the fuel pressure issue, generally requiring less than 3.5 bar. Reciprocating engines that are operating on diesel fuel storage do not have fuel pressure as an issue; however, there may be special permitting requirements for on-site fuel storage of diesel.

Diesel engines should be considered where natural gas is not available or is very expensive. These engines have excellent part-load operating characteristics and high power densities. In most localities, environmental regulations have largely restricted their use for cogeneration. Thus, diesel engines are almost exclusively used for emergency power or where uninterrupted power supply is needed, such as in hospitals and critical data operating centres. As emergency generators, diesel engines can be started and can achieve full power in a relatively short period.

6.6.4   Noise levels

Although fuel cells are relatively expensive to install, they are being tested in a number of sites where typically the cost of a power outage is significant to lost revenues or lost productivity, and where uninterrupted power is mandatory. Their relatively quiet operation has appeal, and these units are being installed in congested commercial areas. Locating a turbine or engine in a residential area usually requires special consideration and design modifications for it to be acceptable.

Engine and turbine installations are often installed in building enclosures in order to attenuate the noise to surrounding communities. Special exhaust silencers or mufflers are typically required on exhaust stacks. Gas turbines require a high volume of combustion air, causing high velocities and associated noise. Inlet air filters can be fitted with silencers to substantially reduce the noise levels. Gas turbines are more easily confined within a factory-supplied enclosure than reciprocating engines.

Reciprocating engines require greater ventilation due to radiated heat that makes their installation in a sound-attenuating building often the most practical solution. Gas turbines require much less ventilation and they can be contained within a compact steel enclosure.

6.6.5   Regulatory and local planning issues

Technical consideration of a cogeneration system must take into account whether the local regulations permit electric utilities to buy electricity from the operators or not. The size and type of the system could be significantly different if the export of electricity to the grid is allowed. Moreover, almost all cogeneration installations will require a planning consent, unless they are contained within an existing site building. This means that various issues such as access, visual impact, noise and construction activity must be addressed. For example, the installation of a new chimney will require authorisation from the local authorities.

In the UK, planning consent is required from the Department of Trade and Industry for gas-fired cogeneration with capacity over 10 MW and for all cogeneration schemes over 50 MW capacity. Moreover, the local environmental regulations can limit the choice of fuel to be used for the proposed cogeneration systems. If the local environmental regulations are stringent, then some of the available fuels cannot be considered because of the high treatment cost of the polluted exhaust gas and, in some cases, of the fuel itself.

Larger cogeneration plants will most probably require authorisation from the corresponding environment agencies regarding emissions and wastes. In particular, large plants installed within urban areas will need to demonstrate that they are not breaching the relevant air quality standards and targets. Plants may also require approval from other regulatory bodies regarding their use of gas as a fuel. In some instances, local community organisations may consider that their views on the installation of a new plant should be considered.

6.6.6   Other issues

Some energy-consuming facilities require very reliable power and/or heat supply. For example, a pulp and paper plant cannot operate with a prolonged unavailability of process steam. In such instances, the cogeneration system that is to be installed must be modular, consisting of more than one unit, so that the unplanned shutdown of a specific unit will not seriously affect the energy supply as a whole. On the other hand, if the cogeneration system is to be installed as a retrofit, then it must be designed such that the existing energy conversion systems, such as boilers, can still be of use.

Dissociation is also needed on whether the cogeneration system will be a grid-dependent or grid-independent one. A grid-dependent system has access to the grid to buy or sell electricity. The grid-independent system is also known as a stand-alone system that meets all the energy demands of the site. It is obvious that for the same energy-consuming facility, the technical configuration of the cogeneration system designed as a grid-dependent system would be different from that of a stand-alone system.

6.7   Practical aspects of installing a cogeneration plant

Although the financial viability of a cogeneration plant is a crucial factor in assessing its feasibility, there are practical aspects that have to be considered even from the initial feasibility study stage. Many of these will be considered in more detail in the later stages of the assessment process, but it is important to ensure that there are no practical obstacles that cannot be overcome as part of the normal cogeneration design and development process or that will probably involve excessive costs.

The location of the cogeneration plant needs to be considered carefully at an early stage, as there are several factors that help to determine its optimum position:

- The plant must be sited where it can remain for a long period without disrupting or obstructing normal site use, either initially or in the future.
- There must be sufficient space to allow access for maintenance and also to house auxiliary equipment.
- The plant must be sited on foundations that are suitable for the static and dynamic loads imposed by it, which may require the construction of a concrete base or, in some cases, the installation of piles.
- The plant must be located in a position from where the recovered heat can be passed into the existing and future site heating systems. This will usually involve installing some new steam or hot water pipework.
- The plant must be connected to the site electrical distribution and fuel supply systems. The availability of such connections merely influences plant location.
- The cogeneration plant may require the installation of a new chimney.
- Although most cogeneration turbines and engines are supplied with acoustic enclosures, the plant and its auxiliary equipment produce noise. Since the plant may operate almost continuously, its location should minimise the impact of the noise emitted.
- The cogeneration plant must be connected to the existing site energy systems, and this may require some modification of utility connections to the site, together with provisions for storing additional fuel, water, lubricants and so on.

6.8   Financial assessment of a cogeneration project

Cogeneration may be considered economical only if the different forms of energy produced have a value higher than the investment and operating costs that are incurred on the cogeneration facility. In some cases, the revenue that is generated from the sale of excess electricity and heat or the cost of availing standby connection must be included. It is harder to quantify the indirect benefits that may accrue from the project, such as the avoidance of economic losses that are associated with disruption in grid power, and the improvement in productivity and product quality.

The major factors that need to be taken into consideration for the economic evaluation of a cogeneration project are the following:

- the required initial investment;
- all operating and maintenance costs;
- the fuel price;
- the price of energy purchased and sold.

The initial investment is a key variable, and this includes many items in addition to the cost of the cogeneration equipment. To start with, one should consider the cost of the pre-engineering and planning. Barring a few exceptional cases, the plant owner/operator would normally hire a consulting firm to carry out the technical feasibility of the project, before identifying the various suitable alternatives that may be retained for economic analysis. If the cogeneration equipment needs to be imported, then the prevailing taxes and duties to the equipment cost should be added.

If the system owner plans to purchase cogeneration components from different suppliers and then assemble them on the site, then the cost of preparing the site for civil, mechanical and electrical works, as well as for acquiring all of the auxiliary items, such as electrical connections, piping of hot and cold utilities, condensers, cooling towers, instrumentation and control, should be taken into account. Table 6.2 provides an example of the breakdown of typical costs for a 20 MWe gas turbine cogeneration plant.

If cogeneration is being adopted as a retrofit at an existing site, then the cost of items will depend greatly on the existing facilities, some of which may be retained while others will be discarded, replaced or upgraded. The cost of land may be a crucial factor at some sites where the cogeneration facility is commissioned, particularly in the case of urban buildings or where additional space is required for storage and handling of fuel.

The integration of the cogeneration plant into the existing set-up may lead to some economic losses to the plant operator through such things as production downtime. The costs that are associated with such losses should be included in the total project cost. The annual costs that are incurred due to the cogeneration plant, such as insurance fees and property taxes, should also be included in the analysis. These are often calculated as a fixed percentage of the initial investment.

Table 6.2 Cost breakdown of a 20 MWe gas turbine cogeneration plant

Component/service type		Costs (US$)
Gas turbine plant equipment		9 000 000
Gas turbine gen-set package	8 100 000	
Auxiliary systems	370 000	
Fuel gas compressor/skid	420 000	
Back-up distillate storage	110 000	
Steam production equipment		2 580 000
Heat recovery boiler with auxiliary firing	1 840 000	
Water treatment system	320 000	
Condenser, feed-water pumps	420 000	
Electrical components		850 000
Substation transformers	320 000	
Switchgear and controls	110 000	
Utility interconnections	420 000	
Services and installation		4 680 000
Engineering design	1 100 000	
Civil works	630 000	
Control and maintenance room	320 000	
Electrical field work	840 000	
Mechanical field work	1 470 000	
Freight and handling	320 000	
Total plant cost		**18 810 000**
Equipment, design and installation	17 110 000	
Contingency (10%)	1 700 000	

The operation and maintenance (O&M) costs should include all of the direct and indirect costs of operating the new cogeneration facility, such as servicing, equipment overhauls and replacement of parts. The cost of employing additional personnel as well as the training that they will need in order to be able to operate the new facility must also be taken into account. Present technology allows for the complete automation of small pre-packaged and pre-engineered units, helping to reduce the O&M costs considerably.

Fuel costs may form the largest component of the operating expenditures. If cogeneration is added to an existing plant, only the fuel cost in excess of that used earlier for the separate heat and power generation may be considered. Since the cogeneration plant is expected to operate for a lengthy period, escalation of the fuel price over time should be included in a realistic manner.

The price of energy bought and sold is a decisive parameter. This includes the net value of electricity or thermal energy that is displaced, as well as any excess electricity or thermal energy sold to the grid. A good understanding of the electric utility's tariff structure is important, which may include energy and capacity charge, time-of-use tariff, standby charges and electricity buy-back rates. As for the fuel, there should be a provision to account for electricity price escalation with time. This is particularly true where utilities depend heavily on fuel in their power generation mix.

6.9   Assessment of financial feasibility

Once the potential operator is satisfied with the rough payback period of a specific cogeneration project, a common and simple procedure of financial feasibility of that particular alternative may be followed, as shown in Figure 6.2.

In the estimation of the net present value (NPV) for a cogeneration project, the total investment costs are taken as cash outflows, and cash inflows are the difference between the annual total cost of the cogeneration system and that of the conventional energy supplies. Sometimes the total discounted costs of different cogeneration alternatives are estimated instead of the NPV of a single alternative. All the cash outflows are considered and discounted to the present value. The option that has the least discounted costs would be selected as the best system.

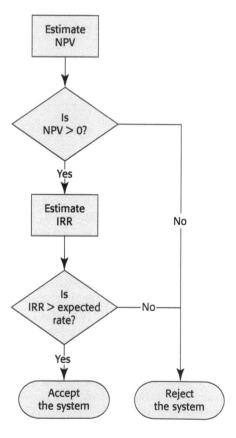

Figure 6.2   Flowchart of cogeneration cost feasibility analysis

Investment decisions are based on these financial indicators that are calculated from cash flow streams. The cash flows are estimated on the basis of a number of factors such as future costs, expected investment levels and tax rates. Therefore, changes in these parameters drastically affect the financial indicators and investment decisions. It is necessary to analyse how the value of a financial indicator (such as internal rate of return) changes when one or more of the input parameters (such as discount rates, fuel prices and investment costs) deviate by a certain amount from the expected value.

This procedure is known as the sensitivity analysis. If the system to be installed has no access to the utility grid, the financial feasibility study will lead to the best cogeneration alternative, since the sizing of different alternatives would have been carried out in the technical feasibility study. Financial indicators are estimated for each cogeneration system retained after the technical feasibility study. The best cogeneration alternative that has the highest NPV, or the least total discounted cost, would be selected.

For systems having access to the utility grid, the optimum size of alternative cogeneration systems is determined by the financial feasibility study. The optimum size of each alternative would be that which has the highest NPV (or least discounted cost). After sizing each alternative, the best alternative that has the highest NPV, or least discounted cost, would be selected.

Normally, the determination of the optimum size of a particular cogeneration system is carried out by computer software, because it is a repetitive and time-consuming process, dealing with a large number of variables and parameters. The objective function of the optimisation process may be the maximisation of the NPV or the minimisation of the total discounted costs. Finally, the best cogeneration system would be identified after the sensitivity analysis is carried out, to make sure that the selected system is still financially attractive with possible variations in the values of some critical parameters.

6.10 Sensitivity analysis

In an initial feasibility study, the calculation of cost savings and installed plant costs is based on estimates of a number of variables. It is important to assess the likely impact of changes in certain variables, as such changes can affect the costs, savings and the payback period. This assessment is called a sensitivity analysis. The variables that need to be included in any sensitivity analysis are as follows:

- cogeneration plant operating hours;
- electricity and heat demands;
- unit prices for fuel and electricity;
- plant installation costs;
- plant maintenance costs.

When considering the impact of a variation in operating hours, it is important to recognise that the cost savings from a cogeneration installation do not vary linearly with annual operating hours, since a disproportionate part of the savings is made during periods when electricity prices are higher. Hourly cost savings vary by a factor of more than 10 between the peak winter electricity tariff periods and the low summer night-time tariff periods. The sensitivity analysis illustrates the differences in cost savings that can result from reduced plant availability at different times of the year.

When considering changes in the costs of fuel and electricity, there are three points of concern:

1. If the costs of electricity increase or decrease, so do the energy cost savings associated with the cogeneration plant.
2. If the costs of fuel increase, the energy savings associated with the cogeneration plant decrease, and vice versa.
3. In terms of cost per unit of energy supplied, electricity is significantly more expensive than primary fuels, and the economics of cogeneration schemes are, therefore, much more sensitive to changes in the unit price of electricity. At current typical prices, a 10 per cent increase in electricity prices improves the payback on a typical cogeneration scheme by 10–15 per cent, while, for comparison, a 10 per cent decrease in fuel prices improves the payback by 3–5 per cent. The combined effect of both changes would be to improve the simple payback by 12–18 per cent.

Chapter 7
Legal and institutional framework

Changing energy policies, such as those in Europe, can influence the profitability of cogeneration plants immensely. In Europe, the introduction of the liberalised electricity market resulted in a fall in electricity prices. This had a negative impact on the profitability of cogeneration plants. On the other hand, the depletion of fossil fuels and environmental concerns has shown the importance of developing energy-saving technologies and energy efficiency measures. Cogeneration presents a potential for substantial increased energy efficiency and a major reduction in environmental impacts. It is therefore of strategic importance.

The EU, along with many of its member states, considers the production of electricity from cogeneration plants to be a priority area. There are several programmes at the community level that support cogeneration. On the national level, the importance of cogeneration has been translated into legal text in many cases. According to a study that has been carried out by ZREU and CRES, nearly 60 per cent of cogeneration experts have expressed a strong interest in learning more about legal and institutional framework conditions. As these change quite often and are sometimes very complex, there is a need for information on this subject.

7.1 European Union law

The EU has promoted the concept of cogeneration since 1974, when an industrial expert group was set up in order to investigate the possibilities of improving upon the conversion efficiency of thermal power stations. In the White Paper 'An Energy Policy for the European Union', the Commission committed itself to presenting a strategy offering a coherent approach for the promotion of cogeneration. New EU initiatives are influencing the future structure and function of Europe's energy industries. This creates a new situation for cogeneration with less price stability for both fossil fuels and electricity, and an increase in environmental concern from the general public.

Cogeneration should play an important role in this new framework. It is therefore of great importance that the efforts to promote cogeneration are consistent with the new industry dynamic. The Commission has published a Communication to the Council and the European Parliament proposing a community strategy to promote cogeneration and to dismantle barriers to its development.

7.2   EU programmes supporting cogeneration

Cogeneration constitutes an important element of the EU CO_2 reduction policies, and an increased share of funding to cogeneration by EU programmes is expected. The Commission has included cogeneration in most of the existing funding programmes, which are discussed further in the text.

7.2.1   Joule–Thermie

The prime objective of the European Commission's Joule–Thermie Programme in the field of non-nuclear energy is to improve energy security by ensuring the provision of durable and reliable energy services at an affordable cost and reducing the impact of the production and use of energy on the environment. Joule–Thermie also aims to contribute to strengthening the technological base of the energy industry.

Thermie's objectives are:

- to improve energy efficiency, in both demand and supply sectors;
- to promote wider use of renewable energy sources;
- to encourage cleaner use of coal and other solid fuels;
- to optimise the use of the EU's oil and gas resources.

The Thermie Programme covers the following sectors:

- rational use of energy in industry, the energy industry;
- fuel cells;
- renewable energy sources;.
- solid fossil fuels (combustion, gasification and waste);
- liquid hydrocarbons.

There is a continued need for the further technological development of the various cogeneration technologies. These developments include improvements to cost-effectiveness, adaptation of new types of application, integration of non-conventional fuel process and improvements of combustion systems to meet tight emission standards. In particular, biomass as a fuel for cogeneration systems deserves specific support. This also applies to the use of new coal cycles and energy-from-waste technologies.

7.2.2   SAVE/ALTENER

These programmes are designed to find solutions to overcome any non-technical barriers that restrict the use of energy efficiency and renewable energy technologies. SAVE has increased the support it gives to actions promoting cogeneration technologies. The ALTENER programme promotes the market penetration of biomass-fired boilers, including cogeneration schemes.

7.2.3   PHARE, TACIS, SYNERGIE and MEDA

The PHARE and TACIS programmes are initiatives provided to encourage cogeneration and sustainable energy developments for the central and eastern European countries, the New Independent States, and Mongolia. They provide support to the process of transformation of these countries to meet the standards of the market economies and to strengthen democracy. Energy is one of the main priorities of TACIS, and it frequently supports cogeneration projects, especially in conjunction with the existing cogeneration-based district heating networks that exist in many of these countries.

The SYNERGIE programme promotes international cooperation in the energy sector and finances various actions promoting cogeneration in Latin America and Asia. This programme can be an important vehicle for the promotion of cogeneration applications in a wide range of Third World countries.

The Euro-Mediterranean partnership (MEDA) is paying particular attention to the energy and environment sectors. Promoting cogeneration through technical assistance and preparatory studies related with district cooling presents an environmental and economic challenge for the countries in this region.

7.2.4   Structural funds

Less favoured European regions can be granted community support for the development of energy efficiency schemes. In Greece, cogeneration is one of the priorities of the operational programme for energy. The European Commission encourages Member States to adopt the development of cogeneration as a priority of national energy programmes financed by the above funds.

7.3   Internalisation of external costs

The internalisation of external costs is a key priority to the integration of environmental considerations into other community policy areas. Energy taxes can act as a stimulus reinforcing cogeneration's competitiveness in the field of electricity and heat production. Cogeneration is an effective and cheap means of improving energy efficiency and reducing pollutant emissions. As such, the principle of internalisation of costs could stimulate an increase in the use of cogeneration technologies.

The European Commission examines ways in which it can integrate the energy and environmental benefits of cogeneration in its taxation policy. Financial instruments such as third-party financing are generally encouraged for cogeneration investments in the industry and the tertiary sector.

The Commission has estimated that there is potential for at least 18 per cent improvement in energy efficiency across the EU simply as a result of market barriers that prevent the satisfactory diffusion of energy-efficient technology and the efficient use of energy. The Commission has instigated an action plan that is intended to remove these barriers.

The SAVE Programme will be used as the principal coordinating arm of the action plan, both as a basis for preparing common action and to provide the means of implementation and evaluation at the community level. Other Community programmes, including the 5th Framework Programme, are also important in this process.

The time frame for the Action Plan covers the period to 2010, and much of its impact will be measurable beyond then. Most of the actions, however, will be initiated during the present lifespan of SAVE and other ongoing Community programmes.

7.4 Other measures

Regional and urban policy and programmes have the potential for large energy efficiency dimensions, should implementing bodies decide to pursue such a course. Energy efficiency can have many positive benefits for the local community, including reducing localised pollution levels, reducing pressure on local transport infrastructure and increasing security of power supplies. The guidelines for the structural funds give priority to the investment by industry in energy-efficient and innovative technologies such as cogeneration.

The directive on common rules for the internal electricity market has had a negative impact on the profitability of cogeneration plants. Nevertheless, the directive takes into consideration the need for environmental protection and energy efficiency measures, and it encourages the use of renewable energy and the production of electricity from cogeneration for the construction of new generating capacity. As for the transmission system operation, a Member State may require the system operator, when dispatching generating installations, to give priority to generating installations using renewable energy sources or waste or producing combined heat and power. This is also valid for the distribution system operation.

7.5 Financial instruments

The lack of available and affordable financing is a critical barrier that manufacturers and other business operators face as they consider cogeneration systems. Plenty of capital is available nationally, but many potential operators have trouble in gaining access to the money that they need at affordable rates. This is especially true for small producers, who often have great difficulty in securing financing, since many of them cannot obtain capital for long-term investments in plant and equipment.

The key issues that companies face when seeking private sector financing are as follows:

- the types of financing available to companies pursuing projects like cogeneration;
- the conventional lenders and their reaction to risk, which affects their view of innovative projects like cogeneration;
- factors in choosing the right lender – and how the lender may 'choose' the borrower;
- lack of basic information on public sector programme types.

There are plenty of private financing sources that are available, ranging from traditional banks to energy services companies (ESCOs) to various vendor-financing schemes. Moreover, a lot of money is available through a number of European and national programmes. Businesses must get a feel for which of these are the most appropriate and they must gain an understanding of how these public and private financing offerings work. They must also be able to show that their needs coincide with programme missions, and that their projects can be shaped to meet lender eligibility requirements and programme award criteria.

Traditional financing is difficult to get. In addition, the normal problems associated with underwriting reviews of loan applications are complicated by several factors:

- lender uncertainty about the viability of proposed process–related changes like cogeneration;
- lender adversity to operations involving new technologies;
- environmental uncertainties that many lenders associate with manu-facturing projects, in terms of lender liability and collateral devaluation.

Financing institutions typically limit their lending to low-risk propositions. There are a number of reasons for this; one of the most important is lenders' concern over how their own regulators will view the viability of their bank operations and lending practices. In practice, this means that lenders are most comfortable with certainty, things they know and processes that they understand. As a result of this, many financing institutions often view innovations or new technologies as 'red flag' situations that are to be avoided in favour of other types of lending.

Many small producers, in fact, are not able to land long-term capital or construction loans at any price; they are viewed as too risky. Their owners often lack enough collateral to meet the underwriting requirements or enough cash to meet the loan processing costs and environmental assessment requirements. While product development initiatives, new technologies and efficiency improvements receive a lot of attention from the public and corporate leaders, bank underwriters often shun them. Innovative projects without a record of success certainty do not often compete well in financial markets because lenders who are looking to their own bottom line are not sufficiently convinced that they will be repaid.

Companies that are seeking financial assistance in order to improve their energy efficiency, production processes and overall competitiveness – including the installation of cogeneration systems – have an array of available financing options to choose from:

- commercial loans
- lease-purchase or vendor financing
- energy services or third-party financing (e.g. through ESCOs)
- retained earnings or company cash flow
- national or EU financial assistance

Businesses must remember that the most appropriate approach will vary according to a number of different factors, such as the size of the operation, the nature of the investment needed, the primary purpose of capital proceeds and the financial health of the company. In each case, the financing options can be divided into two key categories: those that appear on a company's balance sheet and those that do not.

In the first case, the capital purchase of a cogeneration plant will appear on the company's balance sheet as a fixed asset. A capital purchase is generally funded using internal sources, external finance or a mixture of the two. In the second case, two types of organisation can arrange or supply off-balance-sheet financing for the cogeneration plant: equipment supply organisations and energy service company contractors. The usual approach involves an operating lease.

7.6   Internal funding

With internal funding, the company itself provides the necessary capital for the cogeneration plant. As a result of this, it retains full ownership of the project and should reap the maximum potential benefits. However, at the same time, the company bears a considerable element of technical and financial risk, although the degree of this risk can vary with the installation option that has been chosen. For instance, where a company places the work with a turnkey contractor, the contract terms may reduce the risk the company has to bear by placing more of it on the contractor.

Similarly, the terms of contracts with consultants, equipment suppliers and sub-contractors can be designed in order to minimise the investment risk. Internal financing is not necessarily an easy option. One reason for this is that although cogeneration is a long-term investment, it will often have to compete with other potential business projects that are closer to the company's core area activities. Furthermore, the cogeneration scheme may have to compete within a short-term appraisal environment. Thus, obtaining approval for cogeneration as a self-financed project may prove to be a problem.

7.7   Debt finance

A new debt combined with some internal funding often finances a large capital purchase. As with full internal financing, the residual technical and financial risks remain with the investing company, apart from those that specifically lie with suppliers and contractors. At the same time, the company retains the full benefits of the installation. With new debt, it is possible to match an appropriate source of capital to a specific project. In particular, the borrowing timescale can be matched to the timescale of requirements: short-term finance should be obtained for short-term cash needs, and long-term finance for long-term needs, such as cogeneration plant.

For example, if a company that is planning to invest in a cogeneration plant intends to generate a flow of savings/income over a period of 15 years, that company should attempt to finance the plant over the same period. If this is not possible, then the borrowing timescale should, at least, be as long as the payback period for the project plus the period required for recovering the 'cost of money'. In this way, the repayment schedule can be financed out of the savings/income that is generated by the cogeneration system.

7.8   Leasing

Leasing is a financial arrangement that allows a company to use an asset over a fixed period. There are three main types of such an arrangement:

1. hire purchase
2. finance lease (also known as 'lease' or 'full payout lease')
3. operating lease (also known as 'off-balance-sheet' lease)

Under a hire purchase agreement, the purchasing company becomes the legal owner of the equipment once all of the agreed payments have been made. For tax purposes, the company is considered to be the owner of the equipment from the start of the agreement. The basis of the finance lease arrangement is the payment by the company of regular rentals to the leasing organisation over the primary period of the lease. This allows the leasing organisation to recover the full cost – plus charges – of the equipment.

Although the company does not own the equipment, it appears on its balance sheet as a capital item and the company is responsible for all maintenance and insurance procedures. At the end of the primary lease period, either a secondary lease – with much reduced payments – is taken out, or the equipment is sold second-hand to a third party, with the leasing organisation retaining most of the proceeds of the sale.

With finance leasing, the leasing organisation obtains the tax benefits, and these are passed back, in part, to the company in the form of reduced rentals. In principle, the rental can be paid out of the energy savings, thereby assisting the cash flow. Finance leasing may have tax advantages over internal and debt financing if the company has insufficient taxable profits to benefit from the tax allowances that are available on capital expenditure. With this route, the level of financial and technical risk taken on by the company is similar to that of a self-financed project.

7.9   Off-balance-sheet financing options

This option, which is also known as third-party financing (TPF), seems to be an adequate instrument to face financial issues. TPF was developed in order to help companies finance investment without affecting their balance sheets. A user of an efficient and environmentally friendly concept such as cogeneration does not finance the initial outlay for the project. Instead, the operator reimburses the technology supplier by making payments that are related to the performance of the technology that is being installed.

Other forms of TPF include energy services contracts that are provided by ESCOs or utilities, which can offer new services to their customers through cogeneration. There are a wide variety of arrangements that are possible. Under these contracts, an energy service provider agrees with the user on the site for the requirements for heating, lighting, power and so on. It is the responsibility of the contractor to find the most economic method of providing these services, which often involves the installation of a cogeneration plant. This investment is made and managed by the ESCO, which covers it in the charges for the energy services.

The efficiency of cogeneration means that these charges will be lower than the previous site energy costs were. In this scenario, all the sides that are involved with the financial deal profit. Different community programmes can promote this financial scheme by stimulating activities and coordinating interested parties.

7.10   Equipment supplier finance

An equipment supplier may, as an alternative to outright purchase, offer a leasing package to the generator. The equipment supplier will normally design, install, maintain and sometimes operate the cogeneration plant. A common commercial arrangement is for the energy to be supplied at prices that incorporate any agreed discounts on the open market price. The operator pays for the fuel and agrees to buy the electricity and/or the heat generated at the agreed price.

To ensure that the equipment supplier receives a continued income from the sale of electricity and heat energy to the company throughout the 5–10 year contract period, the generator may be required to make a commitment in the form of a substantial standing charge, a lease payment or a high 'take or pay' volume of the energy supplied. This arrangement transfers most of the technical risk from the company to the equipment supplier.

However, the generator's savings are also significantly lower than that under a capital purchase arrangement. The operator also retains the risks that relate to fluctuations in the price of fuel. This form of financing arrangement has commonly been used to finance various small 'packaged' engine-based cogeneration systems.

7.11 ESCO contracts

ESCO arrangements can vary widely. In some instances, the ESCO contractor will design, install, finance, operate and maintain a cogeneration plant on the generator's site. In other cases, the company has the cogeneration plant supplied and installed by one set of contractors under a design or turnkey arrangement, but subcontracts the operation and maintenance of the plant to another contractor. In both of these cases, the ESCO contractor may take on the responsibility for fuel purchase and for other on-site energy plant.

From a financing point of view, the basis of such an agreement is the transfer of cogeneration plant capital and operating costs, together with all the technical and operating risks of cogeneration, from the end user to the ESCO contractor. The generator's savings in this case would normally be less than that under a capital purchase arrangement, because the ESCO contractor needs to recover the cost of the capital investment and cover operating costs, overheads and profit. However, under certain specific circumstances, the savings can be greater than that with a capital purchase arrangement.

For example, the ESCO contractor may be able to size a cogeneration plant that is able to meet the heat requirement of the company and produce surplus electricity that can be exported and sold. The operator will still receive only part of the value of the energy savings, but because the energy savings are greater, the operator's share may turn out to have a value greater than the savings that would have been achieved under a smaller capital purchase scheme. The ESCO contractor will also be able to increase the benefits compared with an in-house solution by avoiding the learning curve costs.

Different ESCO contractors may produce widely differing proposals, depending on the operator's requirements and the ESCO contractor's objectives. Among the many variables to be resolved will be:

- who will operate the plant on a day-to-day basis and, therefore, bear the performance risk?
- who will maintain the plant?
- who will own the plant at the end of the initial agreement period of 10–15 years, and at what on-going cost?

Any transaction with an ESCO contractor still involves a long-term commitment by the producer. Evidence will also be needed in order to satisfy the producer's auditors that the arrangement is an operating lease and not a finance lease. If ownership transfer to the company is implied or stated in the contract, then the arrangement must appear on the company's balance sheet. It should also be noted that an ESCO contract and finance are not necessarily intrinsically linked. It is possible to enjoy the core benefits of an ESCO arrangement, cost reduction and operational risk transfer, irrespective of the finance route that has been chosen.

7.12   Making the choice between options

Choosing an appropriate method of financing will depend on the state of the company's profit/loss account and balance sheet, and also on the degree of risk and benefit that is associated with the project. If a cogeneration system operator opts for a capital purchase, it may obtain the maximum benefits, but will also carry all of the risk. A capital purchase may produce the highest NPV, but the initial cash flow will be negative.

Many producers will not, or cannot, provide the funds for the capital purchase of a cogeneration plant. There may be several reasons for this:

- The return on investment for such a project may be lower than the company's return on capital employed.
- Even if the return on investment is satisfactory, there may be other, more attractive claims on the company's cash resources.
- A capital purchase may increase the company's gearing or reduce liquidity to unacceptable levels.

Therefore, this company may prefer to operate an off-balance-sheet financing option. Where a scheme is financed under an operating lease arrangement, the overall NPV will be lower than it is for the capital purchase option, but the cash flow will always be positive – unless the project is only marginally viable or the lender's charges for money borrowed are high.

ESCOs have shown a lot of ingenuity in devising various schemes that are able to combine the off-balance-sheet advantages of operating leases with the retention of the benefits of capital purchases. However, in recent years, accounting standards have become increasingly strict. It is possible to involve an ESCO contractor with a project, regardless of the financing method that is chosen. Such a company may well have a valuable role to play in managing and lessening the risks to the end user.

7.13   Joint ventures

A number of large-scale cogeneration schemes have been funded as joint ventures between the end user and an ESCO contractor. Joint ventures are a highly specific form of legal entity and they are normally warranted for only large, complex schemes in order to 'ring fence' the operation and to limit the financial liabilities of the partners.

Chapter 8

Future developments

No technology remains static, especially when that technology is still in the early stages of large-scale deployment in many places around the world. Many countries wish to see a major expansion of cogeneration take place, and such a planned expansion – often of an order of magnitude in scale – can result in significant developments of the various technologies used in cogeneration plants. This development can:

- decrease cost
- increase reliability
- increase efficiency
- ease operation and maintenance
- extend life
- ease installation
- improve control
- provide greater interchangeability of parts

8.1 Domestic cogeneration

By far, the biggest potential for massive increases in the level of cogeneration use lies in the domestic market. A number of studies have been carried out that predict that domestic CHP might reach a mass market in Europe by 2010–2015. Some estimates have suggested that by this time, the revenue from domestic cogeneration could reach €2 billion per year.

Domestic cogeneration is a big attraction for governments as it results in a significant reduction in CO_2 emissions, without them having to undertake major expenses to achieve that reduction. In addition to this, the spate of major power outages that occurred in North America and Europe in 2003 resulted in many organisations and consumers paying special attention to new ways of ensuring reliability of supply. Domestic cogeneration offers a possible solution to this concern, especially for users that have a demand that is somewhat larger than the typical domestic consumer, but that is not large enough for them to be regarded as medium-sized consumers. Examples of such consumers might include restaurants, small hotels, retail outlets and community housing.

Any successful domestic cogeneration unit will need to have the following characteristics:

- cheap
- reliable
- easy to install
- easy to maintain
- quiet operation
- easy access to fuel
- low emission levels
- small in size

A number of firms, such as PowerGen and Microgen in the UK, have announced their intention to enter the domestic cogeneration market with units based on a Stirling engine as a prime mover. In the longer term, domestic cogeneration based on fuel cells as the prime mover could become commercially available. However, it is not yet clear which fuel cell technology will end up being dominant in the market. Fuel cells offer a high power-to-heat ratio, and they will thus provide more opportunities to expand the market to include smaller homes with a lower thermal demand, and offer greater opportunities to export excess electricity to the grid.

The Energy Saving Trust (EST) of the UK defines a domestic cogeneration system as one having a maximum output of 3.7 kWe. For these units to be commercially viable, they will need to be small enough to be able to sit in an airing cupboard or under the kitchen sink of an ordinary house. In effect, these systems would be used to replace the main boiler link in an existing heating distribution system.

There is the potential to replace 600 000 new systems in the UK every year, based on the current replacement rate for standard boilers. Over 25 years, this could transform the energy efficiency of 16 million houses in the UK, giving rise to cost savings to consumers of £3.5–4.5 billion per year.

Cambridge Consultants Ltd has predicted that the early adoption of domestic cogeneration could lead to sales and service contracts worth over £1.5 billion per year across Europe by 2010. It also predicts that utilities could see an extra £140 million per year in profits. Ian Halliday-Pegg of Cambridge Consultants said:

With over five million domestic boilers sold every year throughout Europe, domestic cogeneration is a mass volume business opportunity that simply cannot be ignored. It is clear that domestic cogeneration will shake up the home energy market, but there are still fundamental questions that need to be addressed, and much development and cost reduction is required to transform promising technology into viable products and services for the mass market. Market penetration will only be possible when the systems are at a cost that the market can bear.

The UK government has set very ambitious targets for increasing the amount of electricity that is to be generated by cogeneration. Most studies have suggested that these ambitious targets can be met only through the widespread adoption of domestic cogeneration. The 2003 UK Energy White Paper envisaged that the UK energy system will include substantial amounts of domestic cogeneration by 2020. The White Paper predicted that there would be 4 million domestic cogeneration units operating in the UK by 2010. The UK government has launched a number of initiatives to help achieve this target, including:

- a plan within the fuel poverty strategy to install 6 000 domestic cogeneration units in poorer households;
- grants for domestic solar photovoltaic installations;
- policy to address barriers to domestic cogeneration.

The UK government has estimated that of the 24 million households in the UK, 14–18 million of them are suitable for domestic cogeneration units. If domestic cogeneration were to achieve a 100 per cent take-up in this market, it would comprise 20 GWe.

8.2 Advantages of domestic cogeneration

Domestic cogeneration offers substantial advantages:

- A domestic cogeneration initiative could help enable CO_2 emission reduction targets to be met. The current estimates are that the UK alone could save 9–12 million tonnes of CO_2 emissions per year through the use of domestic cogeneration.
- It could help to meet targets for new cogeneration installation. The UK government has set a target of 10 GWe of new cogeneration units to be installed by 2010. The contribution of domestic cogeneration to this target has not yet been quantified, but it has been estimated that replacing 20 per cent of domestic gas-fired boilers in the UK with domestic cogeneration units would provide an installed capacity of about 2.5 GWe.
- It reduces transmission and distribution losses, as domestic cogeneration does not require use of the distribution network. It also helps to reduce the need for reinforcement of the grid.
- The introduction of domestic cogeneration into the electricity market increases the diversity of sources and the flexibility of supply.
- It could enable domestic consumers to take greater advantage of the liberalisation of the energy market. They will be able to sell energy back to the grid.

Domestic customers may enter into a single contract for a bundle of services, including the financing, installation and maintenance of a domestic cogeneration system, along with the installation of other energy-saving measures such as insulation, as well as the sale of gas and electricity.

8.3   Barriers to domestic cogeneration

There are a number of institutional and technical barriers in place that prevent the widespread adoption of domestic cogeneration. At present, customers in the UK can switch suppliers on a 28-day notice, so suppliers are unable to enter into long-term contracts with any certainty. However, the ability to easily switch supplier is key to the liberalisation of the energy market. The DTI is looking at potential ways of enabling this barrier to be overcome.

One of the biggest technical requirements to enable the spread of domestic cogeneration is the development of cheap, reliable two-way meters. These would enable the domestic user to be able to both buy and sell electricity into the grid. This provides security of supply to the domestic user, who will not be without electricity should the domestic unit malfunction, and it also enables the user to offset some of the cost of installing the cogeneration unit by selling excess electricity.

The main technologies that are being considered for use as the prime mover for domestic cogeneration are:

- stirling engine
- fuel cells
- internal combustion engines
- micro-turbines
- others, such as micro-wind turbines or geothermal heat sinks

The UK government has introduced a number of initiatives to encourage micro-CHP development. These include the following:

- The government funded Carbon Trust is running a field trial of micro-CHP units.
- In 2005, the government, subject to the results of field trials, reduced the rate of VAT on micro-CHP units from 17.5 to 5 per cent.

8.4   Domestic cogeneration product status

Plug Power of the USA and Honda of Japan have successfully demonstrated the second generation of their Home Energy Station (HES 2) at Plug Power's corporate headquarters in Latham, New York. Plug Power has also recently introduced the GenSys 5C, its cogeneration fuel cell system, at the Babylon Town Hall in New York state, which is capable of generating 5 kWe and 9 kWth.

MicroGen, which is part of the BG Group, has been developing its MicroGen energy system since 1999. In 2002, the BG Group signed a product development agreement with Rinnai Corporation, Japan's leading boiler manufacturer. To date, BG has invested $30 million in developing domestic cogeneration technology, and it plans to invest another $40 million in order to bring the MicroGen system to launch. The system is designed to generate 1.1 kWe and 5–36 kWth.

In addition, several products are currently being commercially installed in Europe.

8.4.1 SenerTec Dachs

SenerTec Dachs produces 5.5 kWe and 12.5 kWth, and is built around an internal combustion engine as the prime mover. It is manufactured by SenerTec, a subsidiary of Baxi. About 10 000 units have been installed in Germany, and another 100 units have been installed in the Netherlands.

8.4.2 Ecopower

Ecopower is manufactured by Power Plus Technologies, a subsidiary of Valiant. It is built around an internal combustion engine, producing 4.7 kWe and 12.5 kWth. Hundreds of these units have been installed in Germany, along with some in the Netherlands and the UK.

8.4.3 Solo Stirling 161

Solo Stirling 161 is built around a Stirling engine, and can be modified to generate outputs ranging from 2–9.5 kWe to 8–26 kWth.

8.4.4 WhisperGen

In August 2004, E.ON UK announced an order for 80 000 WhisperGen 1 kWe, 8 kWth units. The WhisperGen is based on a Stirling engine, and is manufactured by Whisper Tech. These units have been installed in UK households, where they are undergoing field evaluation.

8.4.5 Others

Other products are undergoing field trials in Europe. These include the following:

- *Sulzer Hexis*: Nearly 100 of Sulzer Hexis' 1 kWe fuel cells have been installed in Germany as part of an extended field trial.
- *Vaillant*: About 20 of Vaillant's 5 kWe fuel cells have been installed as part of the Virtual Fuel Cell Power Plant field trial, partially funded by the EU.
- *RWE Fuel Cells, IdaTech and Buderus*: RWE Fuel Cells, IdaTech and Buderus are working together to field test a small number of 5 kWe fuel cells.
- *Enatec*: Enatec is developing a 1 kWe Stirling engine, and several units are being field tested in the Netherlands.

8.5 Nuclear cogeneration

At the other end of the size scale is nuclear cogeneration. The pebble-bed modular reactor (PBMR) is a high-temperature nuclear reactor that is currently under design in South Africa that will generate significant amounts of usable process heat.

Willem Kriel, the Manager of US Programmes for the PBMR Company, said at a conference in London in January 2006 that the PBMR had a number of commercial opportunities to use the process heat potential to good effect, and that this was an area that the company was seeking to exploit in marketing the reactor. The PBMR is designed to generate 165 MWe and 400 MWth. According to Kriel, the high-temperature heat that is generated by the PBMR can be used for industrial processes, and these could generate an even greater impact than that of electricity generation. These applications include:

- large-scale hydrogen production;
- synthetic natural gas and other liquid and gaseous fuels from coal, oil or other carbon sources;
- process heat for refineries and other chemical plants;
- heat and steam for the recovery of heavy oil and other resources;
- large-scale desalination.

A pilot plant for the PBMR is being constructed in South Africa, and parallel to the work that is being carried out to develop it for electricity generation, work is also being undertaken to prepare for a pilot plant for use in process heat applications, and discussions are underway with a variety of potential industrial users, including the petrochemical industry.

The main drivers in these markets are:

- the cost of natural gas and petroleum;
- energy diversity and security;
- reduction in CO_2 emissions;
- extending the life of carbon resources;
- meeting the growth in energy demand.

The time frame that has been proposed for the development of the PBMR as a process heat provider is:

2005: Business plan developed
2006–07: Project development and planning
2007–12: Technical design and development
2008–12: Regulatory prerequisites
2010–15: Project implemented
2015+: Commercial PBMR process heat plants start operation

8.6 Flexible fuels

Another area that has seen significant development work has been to enable cogeneration plants to be able to make use of a variety of fuels, and to be able to switch from one to another with ease. This is of particular interest to those industrial firms, such as pulp and paper mills, that might have a variety of by-product material that can be burned.

This is also an issue for operators of industrial-size plants that do not operate in cogeneration mode to consider. Cogeneration provides greater flexibility, increases security of fuel supply and enables the plant to take the best advantage of fluctuations in fuel prices. Typical combinations of fuel usage include:

- diesel and gas
- gas and waste by-product
- coal and coal ash
- gas and municipal waste

In particular, municipal waste plants need a firing system that is tolerant of varying fuels. Municipal waste fuel has very variable energy density; that is, the heat energy that can be obtained from burning either a given volume or a given weight of fuel can vary widely. This is because the items that go to make up municipal waste can vary, and clumps of different materials can form, generating fuel with totally different burn characteristics. Furthermore, the moisture and impurity content of the fuel can vary significantly, resulting in uncertainty over how best to reduce or eliminate emissions.

For example, the municipality of Aalborg in Denmark has built a cogeneration plant that is capable of burning biogas produced from a mixture of cattle slurry, organic household waste and organic industrial refuse. The waste is mixed with the cattle slurry and digested under anaerobic conditions at 55 °C in a reactor tank. This generates methane gas, which is used in a cogeneration plant. The degassed and hygienised biomass is returned to the farmers' fields for use as fertiliser.

Another example of a cogeneration plant that burns household waste as a fuel is that of the South East London CHP (SELCHP), which was first established in 1998, in response to the need to find alternative methods of waste disposal, since landfill sites in the area were becoming exhausted. Various options were considered as to how to dispose of the waste without using landfill, and an energy-from-waste scheme was the best option.

The project recoups its investment through four main sources:

1. a gate fee charged for refuse disposal, recognising the avoided cost of landfill;
2. the sale of electricity at the premium price payable for non-fossil fuel sources;

3. the sale of heat to the London Borough of Southwark for use in local homes;

4. the sale of recycled material.

However, because of the way the regulatory system in the UK operates, the sale of heat is being discouraged. This is because the scheme's viability is based on the need to maximise the electricity output. CHP generation, although it is far more efficient in overall fuel use, results in slightly less electricity being generated. Without the heat production element, and converting as much energy as possible into electricity, 40 per cent of the energy content of the waste can be converted to electricity. However, in cogeneration mode, although the overall efficiency of the plant would rise to 85 per cent, only around 30 per cent would be in the form of electricity, with the remaining 55 per cent being heat.

As a result, in addition to the need for technological developments, it is necessary to ensure that the correct financial and regulatory incentives are in place.

8.7 Fuel cells

Fuel cells can be used as the prime mover for cogeneration plants, and they offer the potential for clean, quiet and efficient power generation. Fuel cells are essentially giant chemical batteries that can be charged by a variety of methods, and the energy contained in the fuel cell can be released as required. There are several types of fuel cell, and these are discussed in detail in Chapter 2.

Among new developments in fuel cell technology, Westinghouse has developed its own type of solid oxide fuel cell with tube-shaped cells. The company claims that these can achieve an electrical efficiency of 70 per cent, and that a further 20 per cent can be used for heating.

However, the price of fuel cells is still very high, and this acts as a barrier to the introduction of fuel cells on a large scale. When the mass production of fuel cells starts, the price of fuel cells will probably fall to a more competitive level. However, fuel cells also currently suffer from unproven product durability and reliability.

There will probably be significant developments taking place in the near future in new heat-activated cooling/refrigeration methods, and the use of thermally regenerated dessicants will enhance fuel cell cogeneration applications by increasing the thermal energy loads that is available in certain building types, such as restaurants, supermarkets and refrigerated warehouses.

An example of the potential application in this area can be seen in Japan, where Nuvera Fuel Cells is supplying its Avanti fuel cell power module to the Japan Gas Association. The association is examining the possible development of this fuel cell into water-heating systems that have been developed by the Takagi Industrial Company in order to provide an integrated cogeneration system.

8.8   Trigeneration

Trigeneration is the extension of cogeneration to include cooling. The combination of heating, cooling and power generation offers even greater flexibility for a cogeneration plant. The key to effective cooling is an extra device, an absorption chiller, that generates cooling from the heat that is produced by the cogeneration unit. Absorption chillers can provide very low-cost cooling when they are combined with an appropriate source of waste heat. They can cool at a competitive price using most cogeneration systems.

An absorption chiller is driven by heat. In a conventional chiller – using the mechanical vapour–compression cycle – the refrigerant evaporates at a low pressure in the evaporator, producing a cooling effect. The refrigerant is then compressed in a mechanical compressor to a higher pressure, and it then flows into the condenser. Heat is extracted from here, the refrigerant condenses, and it is then ready to start the cycle over again.

Trigeneration will usually be applied to air conditioning systems for buildings, but it also has potential applications in a number of industrial processes where there can be concurrent and continuous uses for both heating and cooling.

Maximizing the absorption chiller output improves the system efficiency. Cooling of the absorption chiller package and engine system is usually provided by a separate closed-circuit cooling water system that is controlled by an integrated trigeneration control system.

There are additional benefits. Adding an absorption chiller to the system designed for a building allows installation of a cogeneration plant that is larger than the one dictated purely by the heat load, or for the cogeneration plant to run for longer; both of these improve the performance and benefits that are delivered by the plant. The absorption chiller is effectively running on 'waste' heat rather than on the electricity that conventional chillers require.

Trigeneration can, and does, operate a wide range of output sizes, from individual buildings to large-scale district energy schemes, distributing hot water, chilled water and power to many buildings. The best-known UK trigeneration scheme serves a dozen buildings in the city of London, and the technology is also relatively commonly used in the USA. Elsewhere in the world, especially in the Middle East, there are dedicated 'district cooling' facilities that are used to supply chilled water to several buildings from a central plant.

One example of a trigeneration scheme is located in Southampton in England. This scheme is operated by the Southampton Geothermal Heating Company, which is a joint venture between Southampton City Council and its private sector partner. The scheme uses 10 km of buried heating/cooling mains in order to be able to supply heating, power and chilled water to a mixture of commercial and public and domestic buildings. The scheme includes 7 MW of cogeneration plant as well as a renewable geothermal source of heat and a 5.9 MW engine-based cogeneration unit, designed to fuel an absorption chiller for use as summer cooling.

The scheme is unusual in gaining 95 per cent of its revenue from various commercial sector customers, including the local TV headquarters, a super-store and several hotels and hospitals. The annual sales of energy are about 60 GWh.

In absorption cooling, the evaporator and condenser are essentially the same, but a chemical absorber and generator replace the compressor. A pump provides the pressure charge, but as a pump requires much less power than a compressor needs, the electrical power consumption is much lower. The heat input is required to separate the solutions that caused the evaporation, ready for re-use.

The annual sales of large absorption chillers run to 5 000 units in Japan, 3 500 in China, and 500 in the USA. Germany's market is growing steadily, and is increasing by an extra 100 units per year. By contrast, there has been little take-up of the technology in the UK. Only 10–15 large absorption chillers are sold each year in the UK.

There are two types of absorption chiller that are available: Those for chilling to 5 °C and above usually use lithium bromide and water as the solution; units that use water and ammonia can chill down to -50 °C. Both types of absorption chiller require heat input at 80–200 °C, with the higher temperatures generally achieving higher efficiencies.

Development work on absorption chillers has eliminated the problems that they used to suffer with crystallisation effects, where the working fluid in the unit turned solid under certain unusual circumstances. Corrosion effects in absorption chillers have been massively reduced by the development of the right inhibitors.

Absorption plant tends to be more expensive to install per kilowatt than competing conventional chiller plant designs. However, there is development work being carried out to reduce the differential installation cost, and combined with cogeneration systems, the life-cycle economics of absorption plant are usually very attractive. They are ideally suited for applications where there is a significant continuous requirement for both heating and chilling applications.

Seeing significant potential in developments in the USA in this field, the US Department of Energy has awarded several million dollars to allow US manufacturers to design fully compatible system parts for trigeneration use. These awards include:

- United Technologies Research Centre, to design an accelerated trigeneration system based on Capstone's 60 kW micro-turbine coupled to Carrier absorption chillers.
- Burns and McDonnell, teaming with Solar Turbines and Broad USA, to design and construct a trigeneration system for buildings, using power from a Taurus 5.2 MW turbine and 3 000 refrigeration tonnes of absorption chilling.
- Gas Technology Institute, to combine Waukesha engines with Trane absorption chillers. Engine sizes will range from 290 to 770 kW.

Chapter 9

Case studies

9.1 Whitehall District Heating Scheme

The Whitehall District Heating Scheme (WDHS) was first designed way back in the 1930s to replace the inefficient open coal-fire heating of offices that were used at the seat of the UK Government in London. The installation of boilers began in the 1950s, and the full scheme began operating by 1966. Since then, the central boiler plant has had several upgrades in order to improve its performance and economy.

The scheme supplies 23 government buildings through 24 km of distribution pipework that are routed through a complex network of underground tunnels. The buildings that are covered by the scheme include:

- Downing Street
- The Ministry of Defence
- The Foreign and Commonwealth Office
- The Department for Environment, Food and Rural Affairs
- Horse Guards
- The Treasury

The Whitehall District Heating Scheme (WDHS) supplies 33.9 GWh per year. Heat losses from the extensive distribution pipework network come to approximately 2 GWh per year, about 6 per cent of the total heat generated. This figure for heating losses is similar to schemes of the same vintage.

PACE, the owner of the WDHS, decided to carry out a detailed survey of the system in 1993. The survey discovered that while the distribution pipework was still in a satisfactory condition, the central plant was nearing the end of its useful life. As a result of this, PACE needed to carry out a full appraisal of possible replacement systems. The results of this survey indicated that the best course of action was to retain the central heating scheme, but replace the central plant. The next stage was to determine what plant would provide the most cost-effective and reliable source of heat. Cogeneration was an obvious candidate.

A review of the sizes and combinations of possible cogeneration systems showed that the technology would reduce annual energy costs and total net present costs compared with the non-cogeneration option of heat-only boilers. Several funding options were considered for the scheme, including Private Funding Initiative (PFI). However, the capital investment eventually came from the treasury because the scheme was for the government estates.

ELYO (UK) won the contract for the cogeneration plant in December 1995. It employed Parsons Brinckerhoff to design and manage the installation of the plant, for which the capital cost was £7.82 million.

The introduction of cogeneration required the replacement of all of the existing boilers with a combination of cogeneration and new heat-only boilers. The installed system comprised a single 4.9 MWe Alstom Typhoon gas turbine that has a heat output of 8 MW via a waste heat boiler. Four conventional boilers, each rated at 5.8 MW, provided the back-up for the heat supply. Cheap gas supplied on an interruptible basis powered the turbine and boilers. These supplies can be supplemented at peak demand by oil. Experience has shown that the cogeneration system saves £344 000 per year.

The distribution pipework supplies high-temperature hot water at 160 °C to each of the buildings that are in the system. Meters at the point of entry into each building transmit the supply readings to a central control room that processes this data. The data forms the basis for calculating the charges to be made to customers for the heat used. Each building or customer is responsible for its own use of the heat after the heat enters their building.

In the heating season, the cogeneration plant operates as the lead boiler in order to satisfy the demands of the WDHS. The four heat-only boilers are then sequenced on in response to increasing demand. The net electricity that is produced by the turbine is used within the boiler room for pumping needs and to supply the Ministry of Defence building. Any surplus electricity generated by the plant is sold to the grid.

The maintenance of the plant includes essential requirements such as the replacement of the turbine shaft every three years. The maintenance cost of the cogeneration plant averages £130 000 per year, and the cost of maintaining the WDHS distribution system, excluding the generation plant, is £530 000 per year.

9.2 Terra Nitrogen fertiliser plant

The Terra Nitrogen manufacturing plant in Billingham, UK, produces nitrogen for fertilisers, and to do so, it needs a high-volume stable supply of steam. Terra Nitrogen also supplies steam to other chemical manufacturing companies on the Billingham site.

In 1997, the mounting operational costs and the varying reliability of the existing plant persuaded Terra Nitrogen that it needed to build a new energy plant that was capable of generating large volumes of steam at high pressures, so that this plant could provide a rapid delivery of steam on interruption or failure of the site supply.

In 1998, Dalkia signed a contract to design, build, operate and maintain for 15 years a new steam generation facility at the site. The company invested £4 million to cover the construction of the plant. The new energy centre has three 29.5 ton/h boilers. Steam is delivered from these at 350 °C and 20.5 barg, with a requirement for the delivery of steam within 5.5 minutes of loss of pressure on the clients' steam main.

9.3 Fort William paper mill

The UK's Department of Trade and Industry (DTI) provided £5 million from its Bioenergy Capital Grants Scheme to Energy Power Resources Scotland in order to help with the construction of a wood-fired cogeneration plant at Fort William in Scotland for the paper manufacturer Arjo Wiggins.

The new plant replaced the 40-year-old oil-fired generator. In addition to supplying the factory's entire heating requirements, it also provides 80 per cent of its electricity needs, with the spare capacity being sold to the national grid.

9.4 Boots headquarters, Nottingham

Boots has had a long history of generating power at its headquarters and main warehouses located in Nottingham. By the mid-1990s, the boilers on the site were getting old and they needed to be replaced. The old boilers were also failing to meet new emissions criteria.

Different schemes were considered to provide heat and power for Boots' warehouse and headquarters in Nottingham. Any potential scheme had to take into account the age of the building, which had been built in 1928. The best of the options that were considered was to build a new central cogeneration plant located in a new building. The plant began operating in 1996, and has since performed to expectations.

The plant consists of three turbines, all of them being monitored by optimisation software that calculates the running costs and profitability for any number of the turbines operating at any given time. This enables the plant operator to decide whether to turn units on or off, and whether it is more cost-effective to buy power from or sell it to the grid. In addition, the plant has dual-fuel capability.

The installation of the plant has allowed Boots to save an average of £1 million per year. The plant is CHP quality assured and it is exempt from the climate change levy. However, there are concerns about the application of emission trading and the carbon cap. Boots is working with the Carbon Trust and Action Energy in order to reduce carbon emissions. Boot's concern is that if the plant produces as much carbon as it did before, then the company would need to buy carbon credits.

Cost was a major driver in deciding upon this application and optimising its running. In this particular instance, natural gas was chosen for use as the primary fuel. However, this price of natural gas started to rise in 2000, and this particular option probably would not be chosen under current economic conditions.

9.5 New York Presbyterian Hospital

In 1998, the New York Hospital and Presbyterian Hospital merged, thus forming the New York Presbyterian Hospital (NYPH), and becoming the core of an extensive healthcare network. The merger had several goals, including lowering the costs for services through improved efficiencies. One aspect of achieving this goal included the improvement of the energy supply to the hospital premises.

By 2003, greater environmental awareness, climbing energy prices and regulatory actions persuaded NYPH to appoint an Energy Programs Manager who was to manage all of the energy issues at the hospital and improve the efficiency, reliability and power quality as well as reduce energy costs. The North America blackout of August 2003 brought this need into clear focus, and NYPH accepted a plan to install a cogeneration plant that could meet 100 per cent of the energy needs of the inpatient areas of the hospital. It was also decided that the CHP system needed to have a full black-start capability.

NYPH began exploring the various CHP options that would be able to meet its needs of cutting energy costs and increasing the security and the quality of its energy supply. The New York State Energy Research and Development Authority (NYSERDA) provided $50 000 in order to help fund the feasibility study. The study concluded that there was a single project that would be able to meet every goal of the hospital's energy programme, as well as lowering overall plant emissions and saving $5 million a year in energy costs.

In order to help finance the CHP project, NYPH sought to obtain additional NYSERDA funding through the 'CHP Demonstration' programme, and NYSERDA awarded the hospital a $1 million grant to help construct the CHP plant. NYSERDA has so far awarded 32 grants totalling $15.5 million across New York state; NYPH was one of the only two grant recipients to receive the maximum possible award of $1 million.

The plant consists of a gas turbine, duct burner and heat recovery steam generator (HRSG), with a total power output of 7 500 kWe. The HRSG produces around 35 000 lb of high-pressure steam per hour. Supplemental duct firing doubles this steam output to 70 000 lb per hour. The hospital's load profile around the year for both electricity and steam makes the project a particularly attractive one, as the hospital uses steam in the winter for space heating and in the summer to drive a hybrid electric/steam–driven chiller plant.

The seasonal steam peak demand takes place in the summer months, when the hospital requires 6 000 tons of steam-driven cooling (1 refrigeration ton = 12 000 BTU/h of cooling). In addition to this, during the summer, the steam turbine drives the boiler feedwater pumps and the forced-draught fans for air-conditioning loads. The existing boilers, used to meet the summer steam peak of 116 000 kg/h, will be supplemented by the cogeneration plant's steam output.

The plant uses natural gas as a sole fuel source. Although fuel use in the facility will increase by approximately 16 per cent, net emissions of NO_x, CO_2 and SO_x will not exceed the current levels, and the emission density (emissions per unit output) will fall.

The turbine uses dry low-NO_x technology for combustion control that allows for extremely lean combustion.

At the time of writing, the project is in design development. The design of the grid interconnection with the utility Con Edison has been finalised, and all of the necessary environmental permits have been filed. The final selection of the major equipment will be completed by mid-2006 by competitive tender. The CHP plant is due to become fully operational by August 2007.

The increase in demand requires extra capacity, as well as the means of getting that capacity to the demand centre. On-site capacity reduces the need to build distribution infrastructure, and thus CHP applications in New York City are acutely necessary in order to meet the growing power requirements of the biggest load pocket in the USA.

The high cost of electricity in New York makes the economics of cogeneration very attractive. However, to date, there has only been limited development of cogeneration in New York City. The EPA hopes that this project will serve as a successful roadmap on how best to overcome obstacles in such developments, and thus lead to a greater application of cogeneration projects within the city.

9.6 Aberdeen community heating

In 1999, Aberdeen City Council (ACC), which owns about 26 500 properties, adopted a comprehensive plan that it called its Affordable Warmth Strategy. Since then, the ACC has upgraded a large proportion of its housing stock, mainly by improving the heating systems, insulation and building fabric.

The council's main objective was to give tenants affordable warmth and to reduce CO_2 emissions without incurring any significant capital outlay. ACC's initial task was to identify the most cost-effective way of meeting these objectives.

In recent years, the UK Government has started to give active support to residential community heating developments, leading to an increase in grant funding. As a result of this, community heating has become more economically attractive to local authorities.

ACC commissioned and funded a feasibility study into the use of multi-storey community heating with cogeneration. The study was completed in 2000, and it examined the main issues in detail. ACC's main aim in this study was to determine how much funding it could acquire from various sources. The report also examined how feasible a cogeneration scheme would be for a group of seven multi-storey blocks of flats in the Seaton area of the city that were located close together. Their proximity to each other kept the capital expenditure needed for the distribution systems down. They also had the advantage of having warm-air electric heating systems that were due for upgrade. The presence of seven additional blocks of flats nearby allows for the scheme to be extended at a later date.

One recommendation that was made by the report was for the creation of a not-for-profit organisation to finance the heating of ACC's multi-storey housing stock. The consultants produced an original plan for 59 clusters, and this was later reduced to 35 clusters. The report identified one of the 35 clusters – Stokeshill – as being the most appropriate and suitable for a cogeneration scheme. This cluster comprises 288 flats in four multi-storey blocks. The flats in the Stokeshill cluster had electric storage heating that had been installed in the 1970s. The tenure of these properties was 98 per cent council tenants and 2 per cent private owners. Each dwelling had an average National Home Energy Rating of 3.3, which is very poor.

Cogeneration was the most attractive option, particularly when combined with over-cladding of the buildings. However, this would have been prohibitively expensive, so ACC opted for cogeneration only, which would improve the heat rating to 6.0, reduce the tenant heat costs by about 40 per cent and reduce CO_2 emissions by about 40 per cent.

ACC needed to determine how it could finance the scheme, because the initial capital costs of the scheme were high. Without external funding, ACC would only be able to afford to fund one such scheme every 10–12 years, but the ACC had 59 tower blocks in its area of responsibility that needed upgrading. It was recommended that ACC should set up a separate not-for-profit company in order to develop and manage cogeneration schemes across Aberdeen. As a result, Aberdeen Heat and Power (AH&P) was formed and it received a commitment from ACC to provide funding of £215 000 per year to ensure that any bank loan can be repaid. ACC then successfully applied to the Community Energy Programme for grant funding, which is available for up to 40 per cent of the capital costs. This was the first scheme to receive a Community Energy Programme grant.

ACC owns the land on which the energy centre stands. Should AH&P cease to trade, then Scottish law dictates that its assets would revert to the owner of the land.

Having secured the Community Energy Programme funding, applied for planning permission and carried out the full tendering process, ACC was able to proceed with the scheme. An energy centre was built close to one of the four multi-storey blocks. This houses a 210 kWe gas engine cogeneration unit and two 700 kW gas-fired boilers for peak load and back-up.

Each block receives the heat via pre-insulated underground pipes. It is expected that 47 per cent of the electricity that is produced will be sold to dwellings that are served by the heat network. ACC considered using mechanical heat-recovery ventilation units, but rejected them because their installation would cause excessive disruption to the tenants.

The scheme first started to deliver heat in December 2003. ACC learned the following lessons from the project:

- Approach the process strategically.
- Whole-life costing is the best way to establish the real cost and best value.
- External specialist experience is essential.
- The high development workload means that it is advisable to delegate an individual to champion the project and keep it moving.
- An arms-length company arrangement enables acceleration of refurbishment plans.

9.7 Woking Borough Council

Over the last 15 years, Woking Borough Council in the UK has implemented a series of sustainable energy projects. These projects include the country's first small-scale cogeneration heat-fired absorption chiller system, the first local authority private wire residential cogeneration system, the largest domestic photovoltaic cogeneration system in the country, the first large fuel cell cogeneration system and the first public–private joint energy venture services company (ESCO) in the UK.

As a result of these projects, the council has achieved dramatic improvements in energy efficiency and environmental protection in the borough. Energy consumption fell by 40 per cent during the 10-year period between 1990/91 and 2000/01. The council is using the money that is saved by these energy efficiency projects to put into a capital fund, and this fund has been invested in order to provide a revenue stream that will make the energy savings even more cost-effective.

Phase 1 of the sustainable community energy network was the first sustainable community energy project operating in a competitive energy market in the world. The project comprises the civic offices, Victoria Way car park (where the cogeneration plant is located), a new holiday inn hotel that itself has no intrinsic boiler or chiller plant, with these services being supplied by the cogeneration plant, a conference centre and several other hotels, restaurants and night clubs. Any surplus power that is generated by the plant is exported to other local buildings and sheltered housing.

The plant uses distributed cogeneration, large-scale thermal storage, heat-fired absorption cooling, standby and top-up boilers and private wire, heat and chilled water distributed energy system networks. All of the buildings are interconnected with heat mains and high voltage–low voltage private wire networks with a single connection point to the local distribution network at the cogeneration plant.

The Woking project has also included a number of fuel cell and renewable energy cogeneration projects. By integrating these projects into the larger overall green energy project, the unit cost of these smaller projects can be reduced. As a result of this, the council has the largest domestic photovoltaic systems in the UK, and the first such systems to be integrated with cogeneration on private wire networks. Woking now has the largest concentration of solar photovoltaics in the UK, with over 1 MWe of installed capacity. These systems do not transmit their electricity to the grid, but instead they supply customers directly on private wire networks.

9.8   Aylesford Newsprint

Aylesford Newsprint in Kent operates two cogeneration units in order to meet the mill's steam and energy needs. The cogeneration plant, which is a dual-fuel system, partly fired by natural gas and partly by process residue, also supplies other businesses on the site and exports its surplus electricity to the national grid.

The plant has a combustor that is used to burn the process residue, with the resulting heat being used for steam and electricity generation that can be used in the papermaking process. The combustor can burn about 200 dry tonnes of sludge per day. It is a bubbling fluidised bed boiler (BFBB) that generates 26 ton/h of steam.

The cogeneration plant consists of two GE Frame 6 gas turbines, two heat recovery steam generators (HRSGs) and a back-pressure steam turbine. The HRSGs deliver steam at 61 barg and 480 °C.

At maximum capacity, the system produces 93 MW of electrical power and 115 ton/h of steam. The site uses about 66 MW, and therefore typically exports about 27 MW to the national grid.

The cogeneration plant originally received consent in 1992. Commissioning of the first phase of the development took place in 1994. Phase 2 was completed in 1996, and phase 3 in 2001. Commissioning of the phase 3 had cost £15 million, and National Power Cogen installed a second Frame 6 turbine and converted an existing gas-fired boiler into a cogeneration unit, increasing the capacity by 40 MW to the current level of 93 MW. This extension made the site one of the largest industrial cogeneration plants in the UK, with a combined heat and power capacity of greater than 300 MW.

Alan McKendrick, the Chief Executive of Aylesford Newsprint, said: 'The ongoing development of the energy plant underlines the importance of reliable, efficient and cost-effective supplies of electricity and steam.'

Index